D0230438

Praise for *Superior*:

'This is an essential book on an urgent topic by one of our most authoritative science writers'

SATHNAM SANGHERA, author of *The Boy with the Topknot*

'Whether you think of racist science as bad science, evil science, alt-right science, or pseudoscience, why would any contemporary scientist imagine that gross inequality is a fact of nature, rather than of political history? Angela Saini's *Superior* connects the dots, laying bare the history, continuity, and connections of modern racist science, some more subtle than you might think. This is science journalism at its very best!'

JONATHAN MARKS, author of *Tales of the Ex-Apes*

'Angela Saini's investigative and narrative talents shine in *Superior*, her compelling look at racial biases in science past and present. The result is both a crystal-clear understanding of why race science is so flawed, and why science itself is so vulnerable to such deeply troubling fault lines in its approach to the world around us – and to ourselves'

DEBORAH BLUM, author of *The Poison Squad*

'This deeply researched and unsettling book blends history, interviews, and the author's personal experiences growing up as an Indian girl in a white working-class section of London . . . An important and timely reminder that race is "a social construct" with "no basis in biology"' *Kirkus* (starred review)

'Angela Saini's *Superior* is nothing short of a remarkable, brilliant, and erudite exploration of what we believe about the racialized differences among our human bodies. Saini takes readers on a walking tour through science, art, history, geography, nostalgia and personal revelation in order to unpack many of the most urgent debates about human origins, and about the origin myths of racial hierarchies. This beautifully-written book will change the way you see the world'

JONATHAN M. METZL, author of *Dying of Whiteness*

'Some writers have tackled the sordid history of race science previously, but none have gone so deep under the skin of the subject as Angela Saini in *Superior*. In her deceptively relaxed writing style, Saini patiently leads readers through the intellectual minefields of "scientific" racism. She plainly exposes the conscious and unconscious biases that have led even some of our most illustrious scientists astray'

MICHAEL BALTER, science journalist and
author of *The Goddess and the Bull*

By the same author

Inferior:
The true power of women and the science that shows it

Geek Nation:
How Indian science is taking over the world

Superior

The Return of Race Science

Angela Saini

4th ESTATE • *London*

4th Estate
An imprint of HarperCollins*Publishers*
1 London Bridge Street
London SE1 9GF
www.4thEstate.co.uk

First published in Great Britain by 4th Estate in 2019

1 3 5 7 9 8 6 4 2

Copyright © Angela Saini 2019

The right of Angela Saini to be identified as the
author of this work has been asserted by her in accordance
with the Copyright, Design and Patents Act 1988

A catalogue record for this book is available from the British Library

HB ISBN 978-0-00-834100-8
TPB ISBN 978-0-00-829383-3

Printed and bound by CPI Group (UK) Ltd, Croydon, CR0 4YY

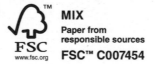

MIX
Paper from
responsible sources
FSC™ C007454

This book is produced from independently certified FSC™ paper
to ensure responsible forest management

Find out more about HarperCollins and the environment at
www.harpercollins.co.uk/green

For my parents,
the only ancestors I need to know.

Contents

Prologue

'In the British Museum is where you can see 'em
The bones of African human beings'
                    – Fun-Da-Mental, 'English Breakfast'

I'M SURROUNDED BY DEAD PEOPLE, asking myself what I am.

Where I am is the British Museum. I've lived in London almost all my life and through the decades I've seen every gallery in the museum many times over. It was the place my husband took me on our first date, and years later, it was the first museum to which I brought my baby son. What draws me back here is the scale, the sheer quantity of artefacts, each seemingly older and more valuable than the last. I feel overwhelmed by it. But as I've learned, if you look carefully, there are secrets – secrets that undermine the grandeur, that offer a different narrative from the one the museum was built to tell.

When medical doctor, collector and slave owner Sir Hans Sloane bequeathed the British Museum's founding collection upon his death in 1753, an institution was established that would come to document the entire span of human culture, in time and space. The British Empire was growing, and in the museum you can still see how these Empire-builders envisioned their position in history. Britain framed itself as the heir to the great civilisations of Egypt, Greece, the Middle East and Rome.

The enormous colonnade at the entrance, completed in 1852, mimics the architecture of ancient Athens. The neo-classical style Londoners associate with this corner of the city owes itself to the fact that the British saw themselves as the cultural and intellectual successors of the Greeks and Romans.

Walk past the statues of Greek gods, their bodies considered the ideal of human physical perfection, and you're witness to this narrative. Walk past the white marble sculptures removed from the Parthenon in Athens even as they crumbled, and you begin to see the museum as a testament to the struggle for domination, for possession of the deep roots of civilisation itself. In 1798, when Napoleon conquered Egypt and a French army engineer uncovered the Rosetta Stone, allowing historians to translate Egyptian hieroglyphs for the first time, this priceless object was claimed for France. A few years after it was found, the British took it as a trophy and brought it here to the museum. They vandalised it with the words, 'Captured in Egypt by the British Army', which you can still see carved into one side. As historian Holger Hoock writes, 'the scale and quantity of the British Museum's collections owe much to the power and reach of the British military and imperial state.'

The museum served one story. Great Britain, this small island nation, had the might to take treasures, eight million exquisite objects from every corner of the globe, and transport them here. The inhabitants of Rapa Nui (Easter Island, as European explorers called it) built the enormous bust of Hoa Hakananai'a to capture the spirit of one of their ancestors, and the Aztecs carved the precious turquoise double-headed serpent as an emblem of authority, but in the nineteenth century both these jewels found their way here and here they've remained. To add insult to injury, they're just two of many, joining objects thousands of years older from Mesopotamia and the Indus

Valley. No single item in the museum is more important than the museum itself. All these jewels brought together like this have an obvious tale to tell, one constructed to remind us of Britain's place in the world. It's a testament to the audacity of power.

And this is why I'm at the museum once again. When I set out to write this book, I wanted to understand the biological facts around race. What does modern scientific evidence really tell us about human variation, and what do our differences mean? I read the genetic and medical literature, I investigated the history of the scientific ideas, I interviewed some of the leading researchers in their fields. What became clear was that biology can't answer this question, at least not fully. The key to understanding the meaning of race is understanding power. When you see how power has shaped the idea of race and continues to shape it, how it affects even the scientific facts, everything finally begins to make sense.

It was not long after the British Museum was founded that European scientists began to define what we now think of as race. In 1795, in the third edition of *On the Natural Varieties of Mankind*, German doctor Johann Friedrich Blumenbach described five human types: Caucasians, Mongolians, Ethiopians, Americans and Malays, elevating Caucasians – his own race – to the status of most beautiful of them all. Being precise, 'Caucasian' refers to people who live in the mountainous Caucasus region between the Black Sea to the west and the Caspian Sea to the east, but under Blumenbach's sweeping definition it encompassed everyone from Europe to India and North Africa. It was hardly scientific, even by the standards of his time, but his vague human taxonomy would nevertheless have lasting consequences. Caucasian is the polite word we still use today to describe white people of European descent.

The moment we were sifted into biological groups, placed in our respective galleries, was the beginning of the madness. Race feels so real and tangible now. We imagine that we know what we are, having forgotten that racial classification was always quite arbitrary. Take the case of Mostafa Hefny, an Egyptian immigrant to the United States who considers himself very firmly and very obviously black. According to the rules laid out by the US government in its 1997 Office of Management and Budget standards on race and ethnicity, people who originate in Europe, the Middle East and North Africa are officially classified as white, in the same way that Blumenbach would have categorised Hefny as Caucasian. So in 1997, aged forty-six, Hefny filed a lawsuit against the United States government to change his official racial classification from white to black. He points to his skin, which is darker than that of some self-identified black Americans. He points to his hair, which is black and curlier than that of some black Americans. To an everyday observer, he's a black man. Yet the authorities insist that he is white. His predicament still hasn't been resolved.

Hefny isn't alone. Much of the world's population falls through some crack or another when it comes to defining race. What we are, this hard measure of identity, so deep that it's woven into our skin and hair, a quality nobody can change, is harder to pin down than we think. My parents are from India, which means I'm variously described as Indian, Asian, or simply 'brown'. But when I grew up in south-east London in the 1990s, those of us who weren't white would often be categorised politically as black. The National Union of Journalists still considers me a 'Black member'. By Blumenbach's definition, being ancestrally north Indian makes me Caucasian. Like Mustafa Hefny then, I too am 'black', 'white' and other colours, depending on what you prefer.

We can draw lines across the world any way we choose, and in the history of race science, people have. What matters isn't where the lines are drawn, but what they mean. The meaning belongs to its time. And in Blumenbach's time, the power hierarchy had white people of European descent sitting at the top. They built their scientific story of the human species around this belief. They were the natural winners, they thought, the inevitable heirs of the great ancient civilisations nearby. They imagined that only Europe could have been the birthplace of modern science, that only the British could have built the railway network in India. Many still imagine that white Europeans have some innate edge, some superior set of genetic qualities that has propelled them to economic domination. They believe, as French President Nicolas Sarkozy said in 2007, that 'the tragedy of Africa is that the African has not fully entered into history . . . there is neither room for human endeavour nor the idea of progress.' The subtext is that history is over, the fittest have survived, and the victors have been decided.

But history is never over. There are objects in the British Museum that scream this truth silently, that betray the secret the museum tries to hide.

When you arrive for the first time it's almost impossible to notice them because they're so easily ignored by visitors in a rush to tick off every major treasure. You join the other fish in the shoal. But go upstairs to the Ancient Egypt galleries, to the plaster cast of a relief from the temple of Beit el-Wali in Lower Nubia, built by the pharaoh Ramesses II, who died in 1213 BCE. It's high near the ceiling, spanning almost the entire room. See the pharaoh depicted as an impressive figure on a chariot, wearing a tall blue headdress and brandishing a bow and arrow, his skin painted burnt ochre. He's ploughing into a legion of Nubians, dressed in leopard skins, some painted with black skin and some the same ochre as him. He sends their limbs into a tangle before they're finally conquered. As the relief shows, the Egyptians at that time believed themselves to be a superior people with the most advanced culture, imposing order on chaos. The racial hierarchy, if that's what you want to call it, looked this way in this time and place.

Then things changed. Downstairs on the ground floor is a granite sphinx from a century or two later, a reminder of the time when the Kushites, inhabitants of an ancient Nubian kingdom located in present-day Sudan, invaded Egypt. There was a new winner now, and the Ram Sphinx protecting King Taharqo – the black king of Egypt – illustrates how this conquering force took Egyptian culture and appropriated it. The Kushites built their own pyramids, the same way that the British would later replicate classical Greek architecture.

Through objects like this you can understand how power balances shift throughout history. They reveal a less simple version of the past, of who we are. And it's one that demands humility, warning us that power is fleeting. More importantly, they show that knowledge is not just an honest account of what we know, but has to be seen as something manipulated by those who happen to hold power when it is written.

The Ancient Egypt galleries of the British Museum are always the most crowded. As we walk past the ancient mummies in their glittering cases we don't always recognise that this is also a mausoleum. We're surrounded by the skeletons of real people who lived in a civilisation no less remarkable than the ones that followed or that went before. Every society that happens to be dominant comes to think of itself as being the best, deep down. The more powerful we become, the more our power begins to be framed as not only cultural but natural. We portray our enemies as ugly foreigners and our subordinates as inferior. We invent hierarchies, give meaning to our own categories. One day, a thousand years forward, in another museum, in another nation, these could be European bones encased in glass, what was once considered an advanced society replaced by a new one. A hundred years is nothing; everything can change within a millennium. No region or people has a claim on superiority.

Race is the counter-argument. Race is at its heart the belief that we are born different, deep inside our bodies, perhaps even in character and intellect, as well as in outward appearance. It's the notion that groups of people have certain innate qualities that are not only visible at the surface of their skins, but are intrinsic to their physical and mental capacities, that perhaps even help define the passage of progress, the success and failure of the nations our ancestors came from.

Notions of superiority and inferiority impact us in deep ways. I was told of an elderly man in Bangalore, south India, who ate his chapatis with a knife and fork because this was how the British ate. When my great-grandfather fought in the First World War for the British Empire and when my grandfather fought in the Second World War, their contributions were forgotten, like those of countless other Indian soldiers. They were considered not strictly equal to their white British counterparts. This is how it was. Generations of people in the twentieth century lived under colonial rule, apartheid and segregation, suffered violent racism and discrimination, because this is how it was. When boys from my school threw rocks at my sister and me when we were little, telling us to go home, this is how it was. I knew even as I bled that this is how it was. This is how it still is for many.

Race, shaped by power, has acquired a power of its own. We have so absorbed our classifications – the trend begun by scientists like Blumenbach – that we happily classify ourselves.

Many of those who visit the British Museum for the first time (I can tell you this from having spent hours watching them) come searching for their own place in these galleries. The Chinese tourists go straight to the Tang dynasty artefacts; the Greeks to the Parthenon marbles. The first time I came here, I made a beeline for the Indian galleries. My parents were born in India, as were their parents, and theirs before them, so this is where I imagined I would find the objects most relevant to my personal history. So many visitors have that same desire to know who their ancestors were, to know what *their* people achieved. We want to see ourselves in the past, forgetting that everything in the museum belongs to us all as human beings. We are each products of it all.

But, of course, that's not the lesson we take, because that's not what the museum was designed to tell us. Trapped inside glass cabinets, fixed to the floors, why are these objects in these rooms, and not where they were first made? Why do they live inside this museum in London, its neo-classical columns stretching into the wet, grey sky? Why are the bones of Africans here, and not where they were buried, in the magnificent tombs that were created for them, where they were supposed to live out eternity?

Because this is how power works. It takes, it claims and it keeps. It makes you believe that this is where they belong. It's designed to put you in your place.

The global power balance, as it played out in the eighteenth century, meant that treasures from all over the world could and would only end up in a museum like this, because Britain was one of the strongest nations at the time. It and other European powers were the latest colonisers, the most recent winners. So they gave themselves the right to take things. They gave themselves the right to document history their way, to define the scientific facts about humankind. European thinkers told us that their cultures were better, that they were the proprietors of thought and reason, and they married this with the notion that they belonged to a superior race. These became our realities.

The truth is something else.

1

Deep Time

Are we one human species, or aren't we?

FLANKING A ROAD dotted with the corpses of unlucky kangaroos, three hundred kilometres inland from the Western Australian city of Perth – and the other end of the world from where I call home – is what feels like a wilderness. Everything is alien to my eyes. Birds I've never seen before make sounds I've never heard. The dead branches of silvery trees, skeleton fingers, extend out of crumbly red soil. Gigantic rocks weathered over billions of years into soft pastel blobs resemble mossy spaceships. I imagine I've been transported to a galaxy beyond time, one in which humans have no place.

Except that inside a dark shelter beneath one undulating boulder are handprints.

Mulka's Cave is one of lots of ancient rock art sites dotted across Australia, but unique in this particular region for being so densely packed with images. I have to crouch to enter, navigating the darkness. One hand is all I see at first, stencilled within a spray of red ochre illuminated on the granite by a diffuse shaft of light. My eyes adjust, more hands appear. Infant hands and adult hands, hands on top of hands, hands all over the ceiling, hundreds of them in reds, yellows, oranges and whites. Becoming clearer in the half-light, it's as though they're pushing through the walls, willing for a high-five. There are parallel lines, too, maybe the vague outline of a dingo.

The images are hard to date. Some may be thousands of years old, others very recent. What is known is that the creation of rock art on this continent goes back to what in cultural terms feels like the dawn of time. Following excavation at the Madjedbebe rock shelter in Arnhem Land in northern Australia in 2017, it was conservatively estimated that modern humans had been present here for around 60,000 years – far longer than members of our species have lived in Europe, and long enough for people here to have witnessed an ice age, as well as the extinction of the giant mammals. And they may have been making art at the outset. At the Madjedbebe site, I'm told by one archaeologist who worked there, researchers found ochre 'crayons' worked down to a nub. At Lake Mungo in New South Wales, a site 42,000 years old, there is evidence of ceremonial burial, bodies sprinkled with ochre pigment that must have been transported there over hundreds of kilometres.

'Something like a handprint is likely to have many different meanings in different societies and even within a society,' says Benjamin Smith, a British-born rock art expert based at the University of Western Australia. It may be to signify place, possibly to assert that someone was here. But meaning is not always simple. The more experts like him have tried to decipher ancient art, wherever it is in the world, the more they've found themselves only scratching at the surface of systems of thought so deep that Western philosophical traditions can't contain them. In Australia, a rock isn't just a rock. The relationship that indigenous communities have with the land, even with inanimate natural

objects, is practically boundless, everyone and everything intertwined.

What looks to me to be an alien wilderness isn't wild at
all. It's a home that is more lived in than any I can imagine.
Countless generations have absorbed and built upon knowledge of food sources and navigation. They have shaped the
landscape sustainably over millennia, built a spiritual relationship with it, with its unique flora and fauna. As I learn slowly,
in Aboriginal Australian thinking, the individual seems to melt
away in the world around them. Time, space and object take
on different dimensions. And none except those who have
grown up immersed in this culture and place can quite understand it. I know that I could spend the rest of my life trying to
fathom this and get no further than I am now, standing lonely
in this cave.

We can't inhabit minds that aren't our own.

I was a teenager before I discovered that my mother might
not actually know her own birth date. We were celebrating her
birthday on the same day in October we always did when she
told us in passing that her sisters thought she had actually been
born in the summer. Pinning down dates hadn't been routine
when she was growing up in India. It surprised me that she
didn't care, and my surprise made her laugh. What mattered
to her instead was her intricate web of family relationships, her
place in society, her fate as mapped in the stars. And so I began
to understand that the things we value are only what we know.
I compare every city I visit with London, where I was born, for
example. It's the centre of my universe.

For archaeologists interpreting the past, deciphering cultures that aren't their own is the challenge. 'Archaeologists have
struggled for a long time to determine what it is, what is that
unique trait, what makes us special,' says Smith, who as well

as working in Australia has spent sixteen years at sites in South Africa. It's a job that has taken him to the cradles of human-kind, rummaging through the remains of the beginning of our species. And this is a difficult business. It's surprisingly tough to date exactly when *Homo sapiens* emerged. Fossils of people who shared our facial features have been found from 300,000 to 100,000 years ago. Evidence of art, or at least the use of ochre, is reliably available in Africa far further back than 100,000 years, before some of our ancestors began venturing out of the con-tinent and slowly populating other parts of the world, including Australia. 'It's one of the things that sets us apart as a species, the ability to make complex art,' he says.

But even if our ancestors were making art a hundred mil-lennia ago, the world then was nothing like the world now. More than forty thousand years ago there weren't just modern humans, *Homo sapiens*, roaming the planet, but also archaic humans, including Neanderthals (sometimes called cavemen because their bones have been found in caves), who lived in Europe and parts of western and central Asia. And there were Denisovans, we now know, whose remains have been found in limestone caves in Siberia, their territory possibly spanning south-east Asia and Papua New Guinea. There were also at vari-ous times in the past many other kinds of human, most of which haven't yet been identified or named.

In the deep past we all shared the planet, even living along-side each other at certain times, in particular places. For some academics, this cosmopolitan moment in our ancient history lies at the heart of what it means to be modern. When we imagine these other humans, it's often as knuckle-dragging thugs. We must have had qualities that they didn't have, something that gave us an edge, the ability to survive and thrive as they went extinct. The word 'Neanderthal' has long been a term of abuse.

Dictionaries define it both as an extinct species of human that lived in ice-age Europe, and an uncivilised, uncouth man of low intelligence. Neanderthals and *Homo erectus* made stone tools like our own species, *Homo sapiens*, Smith explains, but as far as convincing evidence goes, he believes none had the same capacity to think symbolically, to talk in past and future tenses, to produce art quite like our own. These are the things that made us modern, that set us apart.

What separated 'us' from 'them' goes to the core of who we are. But it's not just a question for the past. Today, being human might seem so patently clear, so beyond need for clarification that it's hard to believe that not all that long ago it wasn't so. When archaeologists found fossils of other now-extinct human species in the late nineteenth and early twentieth centuries, they raised doubts about just how far all *Homo sapiens* living today really are the 'same'. Even as recently as the 1960s it wasn't controversial for a scientist to believe that modern humans may have evolved independently in different parts of the world from separate archaic forms. Indeed, some are still plagued by uncertainty over this question. Scientific debate around what makes a modern human a *modern* human is as contentious as it has ever been.

From our vantage point in the twenty-first century, this might sound absurd. The common, mainstream view is that we have shared origins, as described by the 'Out of Africa' hypothesis. Scientific data has confirmed in the last few decades that *Homo sapiens* evolved from a population of people in Africa before some of these people began migrating to the rest of the world around 100,000 years ago and adapting in small ways to their own particular environmental conditions. Within Africa, too, there was adaptation and change depending on where people lived. But overall, modern humans were then (and remain now) one species, *Homo sapiens*. We are special and we are united. It's nothing less than a scientific creed.

But this isn't a view shared universally within academia. It's not even the mainstream belief in certain countries. There are scientists who believe that, rather than modern humans migrating out of Africa relatively recently in evolutionary time, populations on each continent actually emerged into modernity separately from ancestors who lived there as far back as millions of years ago. In other words, different groups of people became human as we know it at different times in different places. A few go so far as to wonder whether, if different populations evolved separately into modern humans, maybe this could explain what we think of today as racial difference. And if that's the case, maybe the differences between 'races' run deeper than we realise.

*

In one early European account of indigenous Australians, the seventeenth-century English pirate and explorer William Dampier called them 'the miserablest people in the world'.

Dampier and the British colonists who followed him to the continent dismissed their new neighbours as savages who had been trapped in cultural stasis since migrating or emerging there, however long ago that was. Cultural researchers Kay Anderson, based at Western Sydney University, and Colin Perrin, an independent scholar, document the initial reaction of Europeans in Australia as one of sheer puzzlement. 'The non-cultivating Aborigine bewildered the early colonists,' they write. They didn't build houses, they didn't have agriculture, they didn't rear livestock. They couldn't figure out why these people, if they were equally human, hadn't 'improved' themselves by adopting these things. Why weren't they more like Europeans?

There was more to this than culture shock. Bewilderment – or rather, an unwillingness to try and understand the continent's original inhabitants – suited Europeans in the eighteenth century because it also served the belief that they were entering a territory they could justly claim for themselves. The landscape was thought to be no different from how it must have been in the beginning, because they couldn't recognise how it might have been changed. And if the land hadn't been cultivated, then by Western legal measures it was *terra nullius*, it didn't belong to anyone.

By the same token, if its inhabitants belonged to the past, to a time before modernity, their days were numbered. 'Indigenous Australians were considered to be primitive, a fossilised stage in human evolution,' I'm told by Billy Griffiths, a young Australian historian who has documented the story of archaeology in his country, challenging the narrative that once painted indigenous peoples as an evolutionary backwater. At least one early explorer even refused to believe they had created the rock art he saw. They were viewed as 'an earlier stage of western history, a living representative of an ancient form, a stepping stone'. From almost the first encounter, Aboriginal Australians were judged to have no history of their own, surviving in isolation as a flashback to how all humans might have lived before some became civilised. In 1958 the late distinguished Australian archaeologist John Mulvaney wrote that Victorians saw Australia as a 'museum of primeval humanity'. Even until the end of the twentieth century writers and scholars routinely called them 'Stone Age' people.

It's true that indigenous cultures have enduring connections to their ancestors, a continuation of traditions that go back millennia. 'The deep past is a living heritage,' Griffiths tells me. For Aboriginal Australians, 'it's something they feel in their bones . . . there are amazing stories of dramatic events that are preserved in oral histories, oral traditions, such as the rising of the seas at the end of the last ice age, and hills becoming islands, the eruption of volcanoes in western Victoria, even meteorites in different times.' But at the same time, this doesn't mean that ways of life have never changed. European colonisers failed to see this and it would take until the second half of the twentieth century for that view to be corrected.

'There was certainly little respect for the remarkable systems of understanding and land management that indigenous Australians had cultivated over millennia,' explains Griffiths. For thousands of years the land has been embedded with stories and songs, cultivated with digging sticks, fire and hand. 'While people have lived in Australia, there's been enormous environmental change as well as social change, political change, cultural change.' Their lives have never been static. In his 2014 book *Dark Emu, Black Seeds*, writer Bruce Pascoe argues, as other scholars have done, that this engagement with the land was so sophisticated and successful, including the harvesting of crops and fish, that it amounted to farming and agriculture.

But whatever they saw, the colonisers didn't value. For those raised in and around cities, industrialisation is still what represents civilisation. 'The idea of ranking, say, an industrial society higher than a hunter-gatherer society is absurd,' reminds Benjamin Smith. It's not easy to accept when you've grown up in a society that tells you concrete skyscrapers are the symbols of advanced culture, but when viewed from the perspective of deep time – across millennia rather than centuries, in the context of long historical trajectories – it becomes clearer. Empires and cities decline and fall. It is smaller, indigenous communities that have survived throughout, those whose societies date to many thousands rather than many hundreds of years. 'Archaeology shows us that all societies are incredibly sophisticated, they are just sophisticated in different ways,' Smith continues. 'These are the world's thinkers, and maybe they thought themselves into a better place. They have societies that have more leisure time than Western societies, lower suicide rates, higher standards of living in many ways, even though they don't have all of the technological sophistication.'

Respect for and pride in indigenous cultures has only started to build in the last few decades. And even now, there remains resistance among some non-indigenous Australians, especially as it has become clear from archaeological evidence that Aboriginal people have been occupying this territory not for just thousands of years, but for many tens of thousands. 'The mid-twentieth century revelation that people were here for that kind of depth of time . . . was received in many ways as a challenge to a settler nation with a very shallow history. There are cultural anxieties wrapped up in all of this,' says Griffiths. 'It challenges the legitimacy of white presence here.'

Among European colonists in the nineteenth century, there was a failure to engage with those they encountered, to accept them as the true inhabitants of the land, combined with a mercenary hastiness to write them off. Alongside the native people of Tierra del Fuego at the southernmost tip of South America, whose nakedness and apparent savagery had shocked biologist Charles Darwin when he saw them on his travels, indigenous Australians and Tasmanians were seen as occupying the lowest rungs in the human racial hierarchy. One observer described them as 'descending to the grave'. They were, Griffiths tells me, seen as doomed to go extinct. 'That was the dominant concept, that they would soon die out.

'There was a lot of talk of smoothing the pillow of a dying race.'

Smoothing the pillow was bloodthirsty work. Disease was the greatest killer, the forerunner of invasion. But starting in September 1794, six years after the First Fleet of British ships arrived in what would become Sydney, and continuing into the twentieth century, hundreds of massacres also helped to slowly and steadily shrink the indigenous population by around 80 per cent, according to some estimates. Many hundreds of thousands of people died, if not of smallpox and other illnesses shipped to Australia, then directly at the hands of individuals or gangs, and at other times of police. Equally harsh was the cultural genocide, adds Griffiths. There were bans on the practice of culture and use of language. 'Many people hid their identity, which also contributed to the decline in population.'

In 1869 the Australian government passed legislation allowing children to be forcibly taken away from their parents, particularly if they were of mixed heritage – described at the time as 'half-caste', 'quarter-caste' and smaller fractions. An official inquiry into the effects of this policy on the indelibly scarred 'Stolen Generations', finally published in 1997, is a catalogue of horrors. In Queensland and Western Australia, people were forced onto government settlements and missions, children removed from about the age of four and placed in dormitories, before being sent off to work at fourteen. 'Indigenous girls who became pregnant were sent back to the mission or dormitory to have their child. The removal process then repeated itself.'

By the 1930s, around half of Queensland's Aboriginal Australian population was living in institutions. Life was bleak, with high rates of illness and malnutrition; behaviour was strictly policed for fear that they would return to the 'immoral' ways of their home communities. Children were able to leave dormitories and missions only to provide cheap labour, the girls as domestic servants and the boys as farm labourers. They were considered mentally unsuited to any other kind of work. Historian Meg Parsons describes what happened as the 'remaking of Aboriginal bodies into suitable subjects and workers for White Queensland'.

Among those forced to live this way were the mother and grandmother of Gail Beck, an indigenous activist in Perth who was once a nurse but now works at the South West Aboriginal Land and Sea Council, fighting to reclaim land rights for her local community, the Noongar. When I visit her at her home in the picturesque port city of Fremantle, speaking to her as she cooks, awaiting a visit from the Aboriginal Australian side of her family, I find someone who has few ways to quantify the pain and loss.

Gail is sixty years old but her true family story is still fairly new to her. Until her thirties, she didn't even know she had any indigenous ancestry. She had been raised to believe she was Italian – a lie to explain her olive skin, her mother terrified that if she were told the truth, Gail might be taken away by the authorities as she herself had been. So she lived under a conspiracy of silence, shielded from the fact that her grandmother had been one of the Stolen Generations, a 'half-caste' taken from her family to live in a Catholic missionary home in 1911 at the age of two. There, she had been abused, physically, mentally, sexually. 'She was put out to service at thirteen. Didn't get paid, nothing like that. And she stayed there until she was an adult.' A similar fate befell Gail's mother, who was under the supervisory care of the nuns in the home from the day she was born, beaten and burned by them when she grew older. The Sisters of Mercy 'were very cruel people', Gail recounts.

Learning about her family's past, and having it confirmed by her grandmother's papers, was a bolt from the blue. 'I cried an ocean of tears.' At once, Gail gained a new identity, one that she was desperate to understand and build a connection to. It took her six years to find the part of her family that had been hidden from her, and she has devoted herself to absorbing their culture ever since. She shows me her blankets and pictures, adorned with the prints for which Aboriginal Australian artists have lately become famous. She has tried to learn an indigenous language, but it has been a struggle. She lives like most white Australians, in a nice house in a nice suburb, her knowledge of her great-grandmother's way of life, as it would have been, fragmentary.

'We are constantly in mourning, and people don't understand that,' she tells me. 'The young children that were lost, that doesn't just affect the nuclear family, that affects the community.' And this is perhaps the greatest tragedy of all, that the way of life she might have had, the knowledge and language she could have been raised with, the relationship to the local environment, all of this was trodden beneath the boot of what considered itself to be a superior race. After the arrival of the Europeans, even the creation of art sharply declined. It took until 1976 for Aboriginal people even to be able to gain legal rights over their land. Throughout, the victims had no choice. 'They weren't allowed to practise their culture, they weren't allowed to mix and they weren't allowed to speak their language.' Having been told they were inferior, that theirs was a life to be ashamed of, they adopted different ways of living – ways they were told were better.

'It was a real shameful thing.'

*

I don't cry easily. But in the car afterwards, I cry for Gail Beck. There is no scale of justice weighty enough to account for what happened. Not just for the abuse and the trauma, the children torn from their parents, the killings, but also for the lives that women and men like her didn't have the chance to live.

In recent decades, as scholars have tried to piece together the past and make sense of what happened, as they share with ordinary Australians in the long process of assessing the damage and its impact, we can see an overarching story about the definition of human difference. It shows us how people have drawn boundaries around other groups of people, and how far inside us and how far back in time the disparities are thought to stretch. These are the parameters of what we now call race.

That same day I meet with Martin Porr, a German-born archaeologist who works at the University of Western Australia, his work focusing on human origins. He feels, as do many archaeologists nowadays, that his is a profession weighed down by the baggage of colonialism. When the first European encounters with Australians happened, when the rules were drawn for how they should be treated, science and archaeology began to be woven in. And they have remained interwoven ever since. For Porr, this tale begins with the Enlightenment, at the birth of Western science. The Enlightenment reinforced the idea of human unity, of an essential biological quality that elevated humans above all other creatures. We live with that concept to this day, seeing it as positive and inclusive, a fact to be celebrated. There was a caveat, however. As Porr cautions, this modern universal way of framing human origins was constructed at a time when the world was a very different place, with far less understanding of other cultures. When European thinkers set the standard for what they considered a modern human, many built it around their own experiences and what they happened to value at that time.

A number of Enlightenment thinkers, including influential German philosophers Immanuel Kant and Georg Wilhelm

Friedrich Hegel, defined humanity without really having much of an idea how most of humanity lived or what it looked like. Those who lived in other lands, including the indigenous people of the New World and Australia, were often a mystery to them. 'A universal understanding of human origins was actually created at the time by white men in Europe who only had indirect access to information about other people in the world through the lens of colonialism,' explains Porr. So when they went out into the real world and encountered people who didn't look like them, who lived in ways they didn't choose to live, the first question they were forced to ask themselves was: Are they the same as *us*?

'If you define humanity in some universal sense, then it's very restrictive. And in the eighteenth century, that was totally Eurocentric. And of course, when you define it in that sense, then of course, so to speak, other people do not meet these standards,' Porr continues. Because of the narrow way Europeans had set their parameters of what constituted a human being, placing themselves as the paradigm, people of other cultures were almost guaranteed not to fit. They didn't necessarily share the same aesthetics, political systems or moral values, let alone food or habits. In universalising humanity, Enlightenment thinkers had inadvertently laid the foundations for dividing it.

And here lay the fatal error at the birth of modern science, one that would persist for centuries, and arguably persists to this day. It is a science of human origins, as British anthropologist Tim Ingold observes, that 'has written the essence of humanity in its own image, and that measures other people by how far they have come in living up to it'.

'When you look at these giants of the eighteenth century, Kant and Hegel, they were terribly racist. They were unbelievably racist!' says Porr. Kant stated in *Observations on the Feeling of the Beautiful and the Sublime* in 1764, 'The Negroes of Africa have by nature no feeling that rises above the trifling.' When he met a quick-witted carpenter, the man was quickly dismissed with the observation that 'this fellow was quite black from head to foot, a clear proof that what he said was stupid.' While a few Enlightenment thinkers did resist the idea of a racial hierarchy, many, including French philosopher Voltaire and Scottish philosopher David Hume, saw no contradiction between the values of liberty and fraternity and their belief that non-whites were innately inferior to whites.

By the nineteenth century, those who didn't live like Europeans were thought to have not yet fully realised their potential as human beings. Even now, notes Porr, when scientists discuss human origins, he still catches them describing *Homo sapiens* in what sound like nineteenth-century European economic terms, as being 'better' and 'faster' than other human species. There's an implicit assumption that higher productivity and more mastery over nature, the presence of settlements and cities, are the marks of human progress, even of evolution. The more we are superior to nature, the more we are superior as humans. It is a way of thinking that forces a ranking of people from closer to nature to more distant, from less developed to more, from worse to better. And history shows us that it's only a small leap from believing in cultural superiority to believing in biological superiority, that a group's achievements are due to their innate capacities.

What Europeans saw as shortcomings in other populations in the early nineteenth century quickly became conflated with how they looked. Cultural scholars Kay Anderson and Colin Perrin explain how, in that century, race came to be *everything*. One writer at the time noted that the natives of Australia differ 'from any other race of men in features, complexion, habits and language'. Their darker skin and different facial features became markers of their separateness, a sign of their permanent difference. Their perceived failure to cultivate the land, to domesticate animals and live in houses was taken as part and parcel of their appearance. And this had wider implications. Race, rather than history, could then be framed as the explanation for not only the Aboriginals' failure, but the failure of all non-white races to live up to the European ideal that Europeans had themselves defined. An Aboriginal Australian – just by having darker skin – could now be lumped together with a West African, for instance, despite being continents apart, with entirely different cultures and histories. Both were black, and this was all that mattered.

Whiteness became the visible measure of human modernity. It was an ideal that went so far as to become enshrined in Australian law. 'When Australia federated in 1901, when the states came together as a nation, one of the first pieces of legislation to pass through Parliament was the Immigration Restriction Act, which formed the basis of the White Australia policy. It sought to fuse the new nation together with whiteness by excluding non-European migration and attempting to assimilate and, ultimately, to eliminate Aboriginal and Torres Strait Islander identity,' explains Billy Griffiths. What happened to Gail Beck's family was one result of these attempts to remove the colour from Australia, in her case to drain it out of her mother's line over generations. 'There was this horrible language of "breeding out the colour" from full-bloods to half-castes to quarter-castes to octoroons,' Griffiths adds. The goal was to steadily replace one 'race' with another.

By the time this state-sanctioned ethnic cleansing was taking place, a crisis had already emerged within scientific circles. Since the Enlightenment, many European thinkers had united around the idea that humankind was one, that we all shared the same common capacities, the same spark of humanity that made it possible for even those of us condemned as 'miserable' to improve, with enough encouragement. Even if there was a racial hierarchy, even if there were lesser humans and greater ones, we were all still *human*. But in the nineteenth century, as Europeans encountered more people in other parts of the world, as they began to see the variety that exists across our species, and failed to 'improve' people the way they wanted to, some began to seriously doubt this cherished belief.

The passage of the nineteenth century saw some make an intellectual shift away from the original Enlightenment view of a single humanity with shared origins. Scientists ventured to wonder whether we all really did belong to the same species.

This wasn't just because of racism. Western scientists had been funnelled into a certain way of thinking about the world partly because of where they happened to be based. In the early days of archaeology, Europe was the reference point for subsequent research elsewhere. Before anyone was sure about humanity's African origins, human fossils in Europe provided the first data. According to John Shea, a professor of anthropology at Stony Brook University in New York, this created an indexing problem. 'If you have a series of observations, the first observations guide you more so than the latter ones. And our first observations about human evolution were based on an archaeological record in Europe.' The first movements out of Africa were eastwards, not westwards. This is why you see elephants in both Africa and Asia. Europe isn't where humans originated – indeed, being so inhospitable back then, it was one of the last places they migrated to, long after going to Australia. But since Europe was where the first archaeologists happened to live and work, this geographical outpost became the model for thinking about the past.

Some of the very oldest human sites in Europe bear evidence of fairly sophisticated cave art. So as a result of indexing, early archaeologists digging on their doorstep logically assumed that art and the ability to think using symbols and images must be a mark of human modernity, one of the features that make us special. But the first *Homo sapiens* arrived in Europe only around 45,000 years ago. When researchers then excavated far earlier sites in Africa, some as old as 200,000 years, they didn't always find the same evidence of symbolic thought and representational art. 'The archaeologists came up with a way to square this,' says Shea. 'They said, well, okay, you know these ancient Africans, Asians, they look morphologically modern but they aren't behaviourally modern. They're not quite right yet.' They decided that although such people *looked* like modern humans, for some reason they didn't *act* like them.

Rather than rethinking what it meant to be a modern human – perhaps taking out the requirement that *Homo sapiens* began making art immediately upon the emergence of our species – the rest of the world's history became a puzzle to be solved. It's a misstep that still has repercussions today. If art is what sets our species apart from Neanderthals and others, then at what point did we actually become our species? Was it 45,000 years ago when we see sophisticated cave art in Europe, or 100,000 years ago when, we now know, people used ochre for drawing? And if Neanderthals or other archaic humans turn out to show evidence of symbolic thought and to have made representational art, will we then have to call them modern too? 'Behavioural modernity is a diagnosis,' says Shea. All the archaeologists can think to do is 'rummage around looking for other evidence that will confirm this diagnosis of modernity'.

In the nineteenth century, such uncertainty around what constituted a modern human being was taken a leap further. If people weren't cultivating the land or living in brick houses, some asked, could they be considered modern? And if they weren't modern, were they even the same species?

Australia in all its alien strangeness posed a particular challenge to European thinkers. Anderson and Perrin argue that the discovery of the continent helped shatter the Enlightenment belief in human unity. After all, here was a remote place, with its own animals not seen elsewhere, kangaroos and koalas, and with its own plants, flowers and unusual landscape. 'Based on observations of the uniqueness of Australian flora and fauna' there were 'suspicions that the entire continent might have been the product of a separate creation,' they write. The humans of Australia were thought to be as strange as everything else there.

After the remains subsequently labelled Neanderthal were first identified in Germany's Neander Valley in 1856, Martin Porr and his colleague Jacqueline Matthews have noted, one of the first things anybody did was compare them to indigenous Australians. Five years later, English biologist Thomas Huxley, a champion of the work of Charles Darwin, described the skulls of Australians as being 'wonderfully near' those of the 'degraded type of the Neanderthal'. It was clear what they were insinuating. If any people on earth were going to have something in common with these now-extinct humans, European scientists assumed, it could only be the strange ones they called savages. Who else could it be but the people who were closest to nature, who had never fitted their definition of what a modern human was?

*

We are forever chasing our origins.

When we can't find what we want in the present, we go back, and back further still, until there at the dawn of time, we imagine we've found it. In the gloomy mists of the past, having squeezed ourselves back into the womb of humanity, we take a good look. Here it is, we say with satisfaction. Here is the root of our difference.

Once upon a time, scientists were convinced that Aboriginal Australians were further down the evolutionary ladder than other humans, perhaps closer to Neanderthals. In 2010 it turned out that Europeans are actually likely to have the largest metaphorical drop of Neanderthal blood. In January 2014 an international team of leading archaeologists, geneticists and anthropologists confirmed that humans outside Africa had bred with Neanderthals. Those of European and Asian ancestry have a very small but tangible presence of this now-extinct human in our lineage, up to 4 per cent of our DNA. People in Asia and Australia also bear traces of another archaic human, the Denisovans. There is likely to have been breeding with other kinds of humans as well. Neanderthals and Denisovans, too, mated with each other. In the deep past, it seems, they were pretty indiscriminate in their sexual partnerships.

'We're more complex than we initially thought,' explains John Shea. 'We initially thought there was either a lot of interbreeding or no interbreeding, and the truth is between those goalposts somewhere.'

The discovery had important consequences. It raked up a controversial, somewhat marginalised scientific theory that had been doing the rounds a few decades earlier. In April 1992 an article had been published in *Scientific American* magazine with the incendiary headline: 'The Multiregional Evolution of Humans'. The authors were Alan Thorne, a celebrated Aus-

tralian anthropologist, who died in 2012, and Milford Wolpoff,
a cheery American anthropologist based at the University of
Michigan, where he still works today. Their hypothesis suggested that there was something deeper to human difference,
that perhaps we hadn't come out of Africa as fully modern
humans after all.

Although this notion had been mooted before, for Wolpoff,
his ideas became cemented in the seventies. 'I travelled and I
looked, I travelled and I looked, I travelled and I looked,' he
tells me. 'And what I noticed was that in different regions, big
regions – Europe, China, Australia, that is what I mean by
regions, not small places – in different regions, it seemed to me
there was a lot of similarity in fossils. They weren't the same and
they all were evolving.'

Wolpoff's big realisation came in 1981 when he was working
with a fossilised skull from Indonesia – one of Australia's closest
neighbours, not far from its north coast – which was dated at
roughly a million years and possibly older. A million years is an
order of magnitude older than modern humans, hundreds of
thousands of years before some of our ancestors first began to
migrate out of Africa. It couldn't possibly be the ancestor of any
living person. Yet Wolpoff says he was struck by the similarities
he thought he could see between its facial structure and that of
modern-day Australians. 'I had reconstructed a fossil that looked
so much like a native Australian to me I almost dropped it,' he
says. 'I propped it up on my lap with the face staring at me . . .
when I turned it over on its side to get a good look at it, I was
really surprised.'

Teaming up with Alan Thorne, who had done related
research and shared his interpretation of the past, they came up
with the theory that *Homo sapiens* evolved not only in Africa,
but that some of the earlier ancestors of our species spread out

of Africa and then independently evolved into modern humans, before mixing and interbreeding with other human groups to create the one single species we recognise today. In their article for *Scientific American*, which helped catapult their multi-regional hypothesis into the mainstream, they wrote, 'some of the features that distinguish major human groups, such as Asians, Australian Aborigines and Europeans, evolved over a long period, roughly where these people are found today.'

They described these populations as 'types', judiciously steering clear of the word 'race'. 'A race in biology is a subspecies,' Wolpoff clarifies when I ask him about it. 'It's a part of a species that lives in its own geographic area, that has its own anatomy, its own morphology, and can integrate with other subspecies at the boundaries . . . There are no subspecies any more. There may have been subspecies in the past – that's something we argue about. But we do know there are no subspecies now.'

Many academics found Wolpoff and Thorne's idea unconvincing or offensive, or both. According to historian Billy Griffiths, the multiregional way of thinking about our origins, undercutting the fundamental belief that we are all human and nothing else, has echoes of an earlier intellectual tradition that viewed 'races' as separate species. 'Wherever we are in the world we look at the deep past and these immense spans of time through the lens of our present moment and our biases and what we want,' he tells me. 'Archaeology is a discipline that is saturated by colonialism, of course. It can't entirely escape its colonial roots.' Multiregionalism, while it was a response to the evidence available at the time, also carried echoes of the politics of colonialism and conquest. 'That's the ugly political legacy that dogs the multiregional hypothesis.'

Wolpoff has always been sensitive to the controversy. He faced down plenty of criticism when he and Alan Thorne published their work. 'We were the enemy,' he recalls. 'If we were right, there couldn't be a single recent origin for humans . . . They said, you're talking about the evolution of human races in separate places independently of each other.'

Their theory remains unproven. Academics in the West and in Africa today generally accept that humans became modern in Africa and then adapted to the environments where they happened to move to fairly recently in evolutionary time – and even these are only superficial adaptations such as skin colour, linked directly to survival. But not everyone everywhere agrees. In China, there's a belief among both the public and leading academics that Chinese ancestry goes back considerably further than the migration out of Africa. One of Wolpoff's collaborators, palaeontologist Wu Xinzhi at the Chinese Academy of Sciences, has argued that fossil evidence supports the notion that *Homo sapiens* evolved separately in China from earlier human species who were living there more than a million years ago, despite data showing that modern Chinese populations carry about as much of a genetic contribution from modern humans who left Africa as other non-African populations do.

'There are many people who are not happy with the idea of African origin,' says Eleanor Scerri, an archaeologist based at the University of Oxford who researches human origins. 'They have co-opted multiregionalism to make a claim that this is a simplistic idea, that races are real, and that people who have come from a particular area have always been there.' She tells me this appears to be prevalent not only in China, but also in Russia. 'There is no acceptance that they were ever African.'

While for some an unwillingness to accept African origins may be motivated by racism or nationalism, it isn't for all. There are those for whom it's simply a way of squaring old origin stories with modern science. In Australia, for instance, Billy

Griffiths tells me, many indigenous people favour the multi-regional hypothesis because it sits closer to their own belief that they have been here from the very beginning. Indeed, this is an origin myth shared by cultures in many parts of the world. Until further evidence comes along (and maybe even after it does), the choice of theory may be driven as much by personal motivations as by data. The past can never be completely known, so the classic multiregional hypothesis persists despite its lack of support among experts. It has political power.

While classic multiregionalism seems unlikely to be the story of our past, the fact that we now know our ancestors bred with other kinds of archaic humans does have implications. It gives nourishment to those who would like to resurrect the multiregional hypothesis in full. It's a factual nugget that feeds fresh speculation about the roots of racial difference. Some dogged supporters of the multiregional hypothesis can rightly claim that at least one prediction made by Wolpoff and Thorne has turned out to be correct. The pair suggested that other now extinct humans such as Neanderthals either evolved into modern humans or interbred with them. And on interbreeding, we now know from genetic evidence, the pair got it right. Some of our ancestors did mate with Neanderthals, although their contribution to people's DNA today runs to just a few per cent, which means it couldn't have been particularly widespread. But it did happen.

When I ask Wolpoff if he feels vindicated by this, he laughs. 'You said vindicated. We said relief!'

Genetics has done the unthinkable, says rock art expert Benjamin Smith. 'The thing that has worried me is the way that genetics research has moved . . . We thought that we were basically all the same, whether you're a bushman in southern Africa, an Aboriginal Australian living in rural Western Australia, or someone like myself who is of European extraction. Everyone was telling us that we were all identical, all the modern science.' The latest discoveries appear to move the story back a little closer to the nineteenth-century account. 'This idea that some of us are more interbred with Neanderthals, some of us are more interbred with Denisovans . . . and Aboriginal Australians had quite a high proportion of Denisovan genetics, for example. That could lead us back to the nasty conclusion that we are all different,' he warns. 'I can see how it might be racialised.'

Indeed, when geneticists revealed the Neanderthal connection, personal ancestry testing companies were quick to sell services offering members of the public the opportunity to find out how much Neanderthal ancestry they might have, using data on genetic variants shared by both humans and Neanderthals – presumably in the expectation that this might mean something to everyday people. Maybe those having the test imagined they would have qualities in common with their extinct cousins.

The finding also had a peculiar effect on scientific research. Fairly soon after it was found to be modern-day Europeans who have the closer association to Neanderthals – not, as it turned out, Aboriginal Australians – the image of the Neanderthal underwent a dramatic makeover. When their remains were first discovered in 1856, the German naturalist Ernst Haeckel had suggested naming them 'Homo stupidus'. But in the twenty-first century, these same Neanderthals, the dictionary definition of simple-minded, loutish, uncivilised thugs, have become oddly rehabilitated.

Svante Pääbo, the director of the genetics department at the Max Planck Institute for Evolutionary Anthropology in Germany, who spearheaded some of the research that led to the discoveries of ancient interbreeding in the first place, was among those to marshal efforts to compare the genomes of

Neanderthals and *Homo sapiens*, in the search for what differs as well as what there is in common. This was accompanied by plenty of speculation from others. In 2018 a set of researchers in Switzerland and Germany suggested that Neanderthals actually had quite 'sophisticated cultural behaviour', prompting one British archaeologist to wonder out loud whether 'they were a lot more refined than previously thought'. An archaeologist in Spain claimed that modern humans and Neanderthals must have been 'cognitively indistinguishable'. A few even raised the possibility that Neanderthals could have been capable of symbolic thought, pointing to freshly discovered cave markings in Spain that appear to predate the arrival of modern humans (the finding has failed to convince rock art expert Benjamin Smith).

'Neanderthals are romanticised,' I'm told by John Shea. They're no longer around, and we don't have a great deal of evidence about what they were like or how they lived, which means they can be whatever we want them to be. 'We're free to project good qualities, things we admire, and the ideal on them.' In reality, whatever they were like, he says, 'the interbreeding thing is more like a symbolic thing for us than it is of evolutionary consequence.'

Yet researchers haven't been able to help themselves looking for evolutionary consequences. One team of scientists claimed that the tiny peppering of Neanderthal DNA may have given Europeans different immune systems from Africans. Another published paper linked Neanderthal DNA to a whole host of human differences, including 'skin tone and hair color, height, sleeping patterns, mood, and smoking status'. An American research group went so far as to try to link the amount of Neanderthal DNA people have with the shapes of their brains, implying that non-Africans may have some mental differences from Africans as a result of their interbreeding ancestors.

For more than a century the word 'Neanderthal' had been synonymous with low intelligence. In the space of a decade, once the genetic link to modern Europeans was suspected and then confirmed, that all changed. In the popular press, there was a flurry of excitement about our hitherto undervalued relatives. Headlines proclaimed that 'we haven't been giving Neanderthals enough credit' (*Popular Science*), that 'they were too smart for their own good' (*Telegraph*), that 'humans didn't outsmart the Neanderthals' (*Washington Post*). Meanwhile a piece in the *New Yorker* whimsically reflected on their apparent everyday similarity to humans, including the finding that they may have suffered from psoriasis. Poor things, they even itched like us. 'With each new discovery, the distance between them and us seems to narrow,' wrote the author. In the popular imagination, the family tree had gained a new member.

In January 2017, the *New York Times* asked: 'Neanderthals were people, too . . . Why did science get them so wrong?' This was indeed the big question. If the definition of 'people' had always included archaic humans, then why should Neanderthals so suddenly be accepted as 'people' now? And not just accepted, but elevated to the celebrity status of sadly deceased genius cousin? It wasn't so long ago that scientists had been reluctant to accept the full humanity even of Aboriginal Australians. Gail Beck's family had been denied their culture, treated in their own nation as unworthy of survival, their children ripped from them to be abused by strangers. In the nineteenth century, *they* had been lumped together with Neanderthals as evolutionary dead-ends, both destined for extinction. But now that kinship had been established between Europeans and Neanderthals, *now* we were all people? *Now* we had found our common ground?

If it had turned out that Aboriginal Australians were the

ones to possess that tiny bit of Neanderthal ancestry instead of Europeans, would our Neanderthal cousins have found themselves quite so remarkably reformed? Would they have been welcomed warmly with such tight hugs? It's hard not to see, in the public and scientific acceptance of Neanderthals as 'people like us', another manifestation of the Enlightenment habit of casting humanity in the European image. In this case, Neanderthals have been drawn into the circle of humankind by virtue of being just a little related to Europeans – forgetting that a century ago, it was their supposed resemblance to indigenous Australians that helped cast actual living human beings out of the circle.

*

Milford Wolpoff is clear with me that he doesn't think there is any biological basis to race, that there are no separate races, except as social categories. He comes across as honest and well meaning, and I believe him. But one obvious implication of his multiregional hypothesis is that if different populations became modern in their own way on their own territories, then maybe some became what we today recognise as human sooner than others. 'A modern human from China looks different than a modern human in Europe, not in the important ways, but in other ways,' he tells me. 'So did one become modern earlier than the other one?' Such a line of thinking opens a door for the politics of today to be projected onto the past, giving rise to racial speculation even if that's not what he intends.

There is still not enough evidence that any humans became modern outside Africa in the way that classic multiregional theory suggests. Even Wolpoff concedes that Africa must remain at the heart of the story. 'I will never say that all of modernity is African, but you've gotta think that most of it is' – even if only because in our deep past that's where most people lived. It is impossible to airbrush Africa out of the lineage of every living person. The genetic evidence we have to date confirms that some version of an 'Out of Africa' scenario must have happened.

But over time, the picture inside Africa has changed to incorporate the growing scientific realisation that our origins might have been a little fuzzier than we imagine. In the summer of 2018 Eleanor Scerri at the Institute of Archaeology in Oxford, together with a large international team of geneticists and anthropologists, published a scientific paper suggesting that rather than humans evolving from a single lineage that can be traced to a single small sub-Saharan African population, perhaps our ancestors were the product of many populations across a far wider area within Africa. These pan-African populations might have been isolated by distance or ecological barriers, and could therefore have been very different from one another. It is multiregionalism, if you like, but within one continent.

'Gradually we started to emerge from the occasional mixing of the populations that were spread around,' Scerri tells me. 'The characteristics that define us as a species don't appear in any single individual until much later. Before that, the characteristics of our species were distributed across the continent in different places at different times.' Modern humans, *Homo sapiens*, emerged from this 'mosaic'. 'We need to look at all of Africa to get a good picture of origins.' This version of our past still puts Africa at its centre, as the first home of our ancestors, but it also concedes that modern humans didn't appear suddenly in one place looking and sounding sophisticated, thinking symbolically and producing art. There was no sudden moment at which the first modern human emerged. The characteristics of us existed in various others before us.

'Humans evolved in Africa first,' agrees anthropologist John Shea. 'Not in just one garden of Eden, but among a broadly distributed population more or less like stops across a subway system. People were moving around along the rivers and coast-lines.' In short, we are a product of longer periods of time and space, a mixture of qualities that incubated in Africa.

According to archaeologist Martin Porr in Australia, this version of the past is more plausible given the way that fossil evidence is scattered across the African continent. For him personally, it also resonates with indigenous Australian ways of defining what it means to be human. Up north in the Kim-berley where he has done most of his work, he says, rock art is not thought of as just images upon rock. 'The rock is actu-ally not a rock but it's a formation out of the dreamtime that is alive, that is in the living world, that people inhabit. And people themselves are part of that.' Human and object, object and environment, are not separated by hard divisions the way they are in Western philosophies.

'You can oscillate in and out of humanity just as objects and animals can oscillate between being human.' An inanimate object can take on human qualities, the way a doll does to a child. In that sense, too, Porr suggests that what made a being human in the past also oscillated.

'I think there's nothing essential about human beings at all.' This, he explains, is how he has come to think about our origins. Not that our evolutionary journey was one big leap, but that we are the gradual products of elements that already existed, in our African ancestors but also in Neanderthals, Denisovans and other archaic humans. Perhaps some of what we think of as purely human characteristics exist in other living creatures today, too.

It's a radically different way of thinking about what it means

to be us, ditching the European Enlightenment view, and taking
a cue instead from other cultures and older systems of thought.
It's a challenge to researchers who have dedicated their careers
to identifying the first modern humans and defining what they
were like, chasing the tail of the Enlightenment philosophers
who thought they already knew. Archaeologists are still trying
to hunt down the earliest cave art, the earliest sign of symbolic,
abstract thought that will signify the leap from a simpler primate
to a sophisticated one, in the hope of pinpointing the magic
moment at which *Homo sapiens* emerged, and where. Geneticists, too, hunt for magical ingredients in our genome, the ones
that will indicate what makes us so remarkable. Increasingly the
evidence suggests that it was never so simple.

'Very few people like looking at human origins from a post-
colonial context, but there is a broader story,' says Porr. There
are other ways of picturing humanity than as a uniquely special
entity far removed from all other living things. Eleanor Scerri
agrees that fresh scientific findings are forcing a rethink of what
it means to be human. 'Popular science needs to get away from
this idea that we originated, and that was *us*. There's never a
time that we were not changing,' she says. 'The idea of these
immutable forms, and that we originate in one place and that's
who we are, that's where we're from.'

What does this mean for us today? If we can't agree on what
makes a modern human, where does that leave the idea of universal humanity? If our origins aren't crystal clear then how do
we know that we're all the same? What does it mean for race?

In a sense, it shouldn't be of any importance. How we choose
to live and treat each other is a political and ethical matter,
one that's already been decided by the fact that as a society we
have chosen to call ourselves human and give every individual
human rights. In reality, though, the tentacles of race reach into

our minds and demand proof. If we are equally human, equally capable and equally modern, then there are those who need convincing before they grant full rights, freedoms and opportunities to those they have historically treated as inferior. They need to be convinced before they will commit to redressing the wrongs of the past, before they agree to affirmative action or decolonisation, before they fully dismantle the structures of race and racism. They're not about to give away their power for free.

And if we're honest, maybe we all need to be convinced. Many of us hold subtle prejudices, unconscious biases and stereotypes that reveal how we suspect we're not quite the same. We cling to race even when we know we shouldn't. A liberal, left-wing British friend of mine, of mixed Pakistani and white English ancestry, who has never been to Pakistan and has no deep ties to the country any more, told me recently that she believes there is something in her blood, something biological within her that makes her Pakistani. I feel this way occasionally about my Indian heritage. But where does culture end and ethnicity begin? Many of us who cherish our ethnic identities, whether on the political left, right or the centre, perhaps betray some commitment to the idea of racial difference.

This is the problem for science. When Enlightenment thinkers looked at the world around them, some took the politics of their day as the starting point. It was the lens through which they viewed all human difference. We do the same today. The facts only temper what we think we already know. Even when we study human origins, we don't actually start at the beginning. We begin at the end, with our assumptions as the basis for inquiry. We need to be persuaded before we cast aside our prior beliefs about who we are. The way new research is interpreted is always at the mercy of the old ideas.

'You can either use the present to explain the past. Or you can use the past to explain the present,' John Shea tells me. 'But you can't do both.' To make sense of the past – and of ourselves – is not a simple job of gathering together scientific data until we have the truth. It isn't just about how many fossils we have or how much genetic evidence. It's also about squaring the stories we have about who we are with the information we're given. Sometimes this information becomes slotted into the old stories, reinforcing them and giving them strength, even if it needs to be forced like a square peg into a round hole. Other times, we have to face the uncomfortable realisation that a story must be ditched and rewritten because however hard we try it no longer makes sense.

But the stories we're raised on, the tales, myths, legends, beliefs, even the old scientific orthodoxies, are how we frame everything we learn. The stories are our culture. They are the minds we inhabit. And that's where we have to start.

2

It's a Small World

How did scientists enter the story of race?

ONCE, A LONG TIME AGO, I floated around the earth in the space of minutes.

I was on a ride at the Magic Kingdom in Walt Disney World, Florida, my little sisters and I perched alongside each other in a slow mechanical boat, buoyed by sugar. 'It's a Small World (After All)' chimed in tinny children's voices, while minuscule automata played out cultural stereotypes from different countries. From what I can recall, there were spinning Mexicans in sombreros and a ring of African dancers laughing alongside jungle animals. Indian dolls rocked their heads from side to side in front of the Taj Mahal. We sailed past, given just enough time to recognise each cultural stereotype, but not quite enough to take offence.

This long-forgotten vignette from my childhood is what comes back to me on the drizzly day I approach the eastern corner of the Bois de Vincennes woodland in Paris. I had heard that somewhere here I'd find the ruins of a set of enclosures in which humans were once kept – not as cruel punishment by the authorities, and not by some murderous psychopath. Apparently they were just ordinary, everyday people, kept here by everyday people, for the fascination of millions of other everyday people, for no other reason than where they happened to come from and what they happened to look like.

'Man is an animal suspended in webs of significance he himself has spun,' American anthropologist Clifford Geertz wrote in 1973. These webs are ours only until someone comes along to pull at the threads. The nineteenth century had marked an age of unprecedented movement and cultural contact, turning the world into a smaller place than it had ever been. It was less mysterious, perhaps, but no less fascinating. And people wanted to see it all. So in 1907 there was a grand Colonial Exposition on this overgrown site in Paris, within the Bois, in what was known as the Garden for Tropical Agriculture, recreating the different parts of the world in which France had its colonies.

Eight years earlier, the garden had been founded as a scientific project to see how crops in distant lands might be better cultivated, helping to bring in more income for colonisers back in Europe. This exposition went a step further. To exotic plants and flowers it added people, displaying them in houses vaguely typical of the ones they might have left behind, or at least how the French imagined them to be. There were five mini 'villages' in all, each designed to be as realistic as possible so visitors could experience what normal life was like for these foreigners. It was an Edwardian Disneyland, not with little dolls, but actual people. They transformed the tropical garden into nothing less than a human zoo.

'In Paris, there were many exhibitions with human zoos,' says French anthropologist Gilles Boëtsch, former president of the scientific council at the National Center for Scientific Research, who has studied their dark history. There was a circus element to it all, a cultural extravaganza. But there was also a genuine desire to showcase human diversity, to give a glimpse of life in the faraway colonies. According to some estimates, the 1907 Paris Exposition attracted two million visitors in the space of just six months – a hit with curious citizens who wanted to see the world in their backyard.

Wherever they were held, most evidence of human zoos has long disappeared, most likely deliberately forgotten. The Garden for Tropical Agriculture is one rare exception. That said, the French authorities don't appear to want to brag about it. It's tucked behind some quiet and well-to-do apartment blocks with barely any signposting. Greeting me as I enter is a Chinese arch that was once probably bright red, but has since faded to a dusty grey. As I walk under it down a gravel path, the place is peaceful but dilapidated. To my surprise, most of the buildings have survived the last century fairly intact, as though everything was abandoned immediately after the tourists left.

To one side is a weathered sculpture of a naked woman, reclining and covered in beads, her head gone, if it was ever there at all. A solitary jogger runs past.

For European scientists, zoos like this offered more than fleeting amusement value. They were a source of biological data, a laboratory stocked with captive human guinea pigs. 'They came to the human zoos to learn about the world,' explains Boëtsch. Escaping the bother of long sea voyages to the tropics, anatomists and anthropologists could conveniently pop down to their local colonial exhibition and sample from a selection of cultures in one place. Researchers measured head size, height, weight, colour of skin and eyes, and recorded the food these people ate, documenting their observations in dozens of scientific articles. With their notebooks, they set the parameters for modern race science.

Race itself was a fairly new idea. Some of the first known uses of the word date from as recently as the sixteenth century, but not in the way we use it now. Instead, at that time it referred to a group of people from common stock, like a family, a tribe, or perhaps – at a long stretch – a small nation. Even until the European Enlightenment in the eighteenth century, many still thought about physical difference as a permeable, shifting quantity. It was rooted in geography, perhaps explaining why people in hotter regions had darker skins. If those same people happened to move somewhere colder, it was assumed their skins would automatically lighten. A person could shift their identity by moving place or converting to another religion.

The notion that race was hard and fixed, a feature that people couldn't choose, an essence passed down to their children, came slowly, and in large part from Enlightenment science. Eighteenth-century Swedish botanist Carl Linnaeus, famous for classifying the natural world from the tiniest insects to the biggest beasts, turned his eye to humans. If flowers could be sorted by colour and shape, then perhaps we too could fall into groups. In the tenth edition of *Systema Naturae*, a catalogue published in 1758, he laid out the categories we still use today. He listed four main flavours of human, respectively corresponding to the Americas, Europe, Asia and Africa, and each easy to spot by their colours: red, white, yellow and black.

Categorising humans became a never-ending business. Every gentleman scholar (and they were almost exclusively men) drew up his own dividing lines, some going with as few as a couple of races, others with dozens or more. Many never saw the people they were describing, instead relying on second-hand accounts from travellers, or just hearsay. Linnaeus himself included two separate sub-categories within his *Systema Naturae* for monster-like and feral humans. However the lines were drawn, once defined, these 'races' rapidly became slotted into hierarchies based on the politics of the time, character conflated with appearance, political circumstance becoming biological fact. Linnaeus, for instance, described indigenous Americans (his 'red' race) as having straight black hair and wide nostrils, but also as 'subjugated', as though subjugation were in their nature.

And so it began. By the time human zoos were a popular attraction, when the ghostly enclosures of the Bois de Vincennes were not eerily empty as they are now, but full of performers – when I would have more likely been within a cage than outside it – the parameters of human difference had become hardened into what we recognise them as today.

Paris wasn't the only city to enjoy this breed of spectacle. Other European colonial powers hosted similar events. Indeed by the time of the 1907 Paris Exposition, human zoos had been around for more than a century. In 1853 a troupe of Zulus undertook a grand tour of Europe. And forty-three years before this an advertisement in London's *Morning Post* newspaper signalled the arrival of a woman who would go down in history as one of the most notorious of all racial freak shows, her story echoed by those to come. 'From the Banks of the River Gamtoos, on the Borders of Kaffraria, in the interior of South Africa, a most correct and perfect Specimen of that race of people,' it announced.

The 'Hottentot Venus', as she was described in the paper, was available for anyone to take a peek at, for a limited time only and at the cost of two shillings. Her real name was Saartjie Baartman and she was aged somewhere between twenty and thirty. What made her so fascinating were her enormous buttocks and elongated labia, considered by Europeans to be sexually grotesque. Calling her a 'Venus' was a joke at her expense. The *Morning Post* took pains to mention the expense shouldered by Boer farmer Hendric Cezar in transporting her all the way to Europe. He was banking on her body causing a scandal.

Baartman had been Cezar's servant in Africa, and by all accounts, she had come with him to Europe of her own free will. But it's unlikely that the life she endured as his travelling exhibit was what she expected. Her career was brief and humiliating. At each show, she was brought out of a cage to parade in front of visitors, who poked and pinched to check that she was real. Commentators in the press couldn't help but notice how unhappy she seemed, even remarking that if she felt ill or unwilling to perform, she was physically threatened. To add to the humiliation, she became, quite literally, the butt of jokes across the city, rendered in relentless caricature like a Twitter meme of Kim Kardashian.

At the end of her run, Baartman ended up in Paris. She found herself at the mercy of celebrated French naturalist Georges Cuvier, a pioneer in the field of comparative anatomy, which aims to understand the physical differences between species. Like so many before him, he was spellbound by her – but his was an anatomist's fascination, one that drove him to undertake a detailed study of every bit of her body. When she died in 1815, just five years after being displayed in London, Cuvier dissected her, removing her brain and genitals and presenting them in jars to the French Academy of Sciences.

As far as Cuvier was concerned, this was just science and she was just another sample. The prodding, cutting, dehumanising fingers of researchers like him sought only to understand what made her and those like her different. What gave some of us dark skin and others light? Why did we have different hair, body shape, habits and language? If we were all one species, then why didn't we look and behave the same way? These were questions that had been asked before, but it was nineteenth-century scientists who turned the study of humans into the most gruesome art. People became objects, grouped together like museum exhibits. Any sense of common humanity was left at the door, replaced by the cold, hard tools of dissection and categorisation.

Following a lifetime of being relentlessly poked and prodded, Baartman remained on show for a hundred and fifty years after her death. Her abused body ended up at the Musée de l'Homme, the Museum of Man, looking out on the Eiffel Tower, a plaster cast of it still standing there until as recently as 1982. It was only in 2002, after a request from Nelson Mandela, that her remains were removed from Paris and finally returned to South Africa for burial.

*

'In the modern world we look to science as a rationalisation of political ideas,' I'm told by Jonathan Marks, a genial, generous professor of anthropology at the University of North Carolina at Charlotte. He is one of the most outspoken voices against scientific racism. Race science, he explains, emerged 'in the context of colonial political ideologies, of oppression and exploitation. It was a need to classify people, make them as homogeneous as possible.' By grouping people and dividing these groups, it was easier to control them.

It is no accident that modern ideas of race were formed during the heyday of European colonialism, when those in power had already decided on their superiority. By the nineteenth century, the possibility that races existed and some were inferior to others gave colonialism a moral kick in the drive for public support. The truth – that European nations were motivated by economic greed or power – was harder to swallow than the suggestion that the places they were colonising were too uncivilised and barbaric to matter, or that they were actually doing the savages a favour.

In the United States, the same tortured logic was used to justify slavery. The transatlantic trade in slaves officially ended in 1807 once the United Kingdom passed its Slave Trade Act, but the exploitation continued for far longer. The use of slave labour continued, people's bodies plundered both in life and death. Dead black slaves, for instance, were routinely stolen or sold for medical dissection. Daina Ramey Berry, professor of history at the University of Texas at Austin, has documented the economic value of slavery in the United States. She notes that there was a brisk trade in black corpses in the nineteenth century, some exhumed by their owners for a quick profit. It's ironic that much of our modern scientific understanding of human anatomy was built on the bodies of those who were considered at the time less than human.

'If you could say that the slavers were naturally distinct from the slaves, then you have essentially a moral argument in favour of slavery,' explains Jonathan Marks. Given this distinction, many feared that the abolition of slavery would set free the human zoo, unleashing chaos. In 1822 a group calling itself the American Colonisation Society bought land in West Africa to establish a colony named Liberia, now the Republic of Liberia, motivated largely by the desperate dread that freed black slaves would want to settle among them, with the same rights. Repatriation to the continent of their ancestors seemed like a convenient solution, ignoring the fact that after generations in slavery, most black Americans simply didn't have a tangible connection to it any more – let alone to a new country that their ancestors may never have seen.

Louis Agassiz, a Swiss naturalist who had been mentored by Georges Cuvier and moved to America in 1846, argued passionately against blacks being treated the same as whites. Shaken by such an intense physical disgust towards black domestic workers serving him food at a hotel that he almost couldn't eat there at all, he became convinced that separate races originated in different places, with different characters and intellectual abilities.

Enslavement was turned back on the slaves themselves. They were in this miserable, degrading position not because they had been forcibly enslaved, it was argued, but because it was their biological place in the universe. At a meeting of the British Association for the Advancement of Science in Plymouth in 1841, an American slave owner from Kentucky named Charles Caldwell had already claimed that Africans bore more of a resemblance to apes. In their 1854 book *Types of Mankind*, American physician Josiah Clark Nott and Egyptologist George Gliddon went so far as to sketch actual comparisons between the skulls of white and black people, alongside those of apes. While the typical European face was artfully modelled on classical sculpture, African faces were crude cartoons, exaggerating features that made it seem they had more in common with chimpanzees and gorillas.

Propelled by a belief that black people had their own unique diseases, Samuel Cartwright, a medical doctor practising in Louisiana and Mississippi, characterised in 1851 what he saw as a mental condition particular to black slaves, coining it 'drapetomania', or 'the disease causing Negroes to run away'. Harvard University historian Evelynn Hammonds, who teaches Cartwright's story to her students, laughs darkly when she recounts it. 'It makes sense to him, because if the natural state of the negro is to be a slave, then running away is going against their natural state. And therefore it's a disease.'

For Hammonds, another chilling aspect of Cartwright's work is the way in which he methodically described the sufferers of drapetomania. 'The colour of the skin is the main difference,' she reads for me from her notes, '. . . the membranes, the muscles, the tendons, all fluids and secretions, then the nerves, and the bile. There's a difference in the flesh. The bones are whiter and harder, the neck is shorter and more oblique.' Cartwright continues this way, couching racism in medical terminology. 'These kinds of observations turned into questions to be explored going forward. Since the 1850s, people have been trying to figure out if black bones are harder than white bones,' Hammonds explains. Cartwright's medical 'discoveries' were patently rooted in the desire to keep slaves enslaved, to maintain the status quo in the American South where he lived. In place of universal humanity came a self-serving version of the human story, in which racial difference became an excuse for treating people differently. Time and again, science provided the intellectual authority for racism, just as it had helped define race to begin with.

Race science became a pastime for non-scientists, too. French aristocrat and writer Count Arthur de Gobineau, in *An Essay on the Inequality of the Human Races*, published in 1853, proposed that there were three races, with what he saw as an obvious hierarchy between them: 'The negroid variety is the lowest, and stands at the foot of the ladder . . . His intellect will always move within a very narrow circle.' Pointing to the 'triangular' face shape of the 'yellow race', he explained that this was the opposite of the negroid variety. 'The yellow man has little physical energy, and is inclined to apathy . . . He tends to mediocrity in everything.' Neither could be a match for Gobineau's own race.

Reaching his predictable pinnacle, Gobineau added, 'We

come now to the white peoples. These are gifted with reflect-
ive energy, or rather with an energetic intelligence. They have
a feeling for utility, but in a sense far wider and higher, more
courageous and ideal, than the yellow races.' His work was a
naked attempt to justify why those like him deserved the power
and wealth they already had. This was the natural order of
things, he argued. He didn't need hard evidence for his theories
because there were plenty of people around him ready and willing to agree that they, too, belonged to a superior race.

It would be Gobineau's ideas that would later help reinforce
the myth of racial purity and the creed of white supremacy. 'If
the three great types had remained strictly separate, the supremacy would no doubt have always been in the hands of the finest
of the white races, and the yellow and black varieties would
have crawled forever at the feet of the lowest of the whites,' he
wrote, promoting a notion of an imaginary 'Aryan' race. These
glorious Aryans, he believed, had existed in India many centuries ago, speaking an ancestral Indo-European language, and
had since spread across parts of the world, diluting their superior bloodline.

Myth and science coexisted, and both served politics. In the
run-up to the passage in 1865 of the 13th Amendment, abol-
ishing slavery in the United States, the race question wasn't
resolved – it just became thornier. Although many Americans
believed in emancipation on moral grounds, fewer were convinced that full equality would ever be possible, for the simple
reason that groups weren't biologically the same. Even Presi-
dents Thomas Jefferson and Abraham Lincoln believed that
blacks were inherently inferior to whites. Jefferson, himself a
slaveholder, agreed with those who thought that the best way to
deal with freed slaves was to send them to a colony of their own.
Freedom was framed as a gift bestowed on unfortunate black

slaves by morally superior white leaders, rather than a reflection of a hope that everyone would one day live alongside each other as friends, colleagues and partners.

*

Not all scientists were quite so self-serving. For those who wanted to establish the facts about human difference, there were unanswered questions. The biggest puzzle was that there was no fleshed-out mechanism to account for how different races – if they were real – might have emerged. If each race was distinct, then where did they each come from, and why? Going by the Bible, as many Europeans did, one explanation for the existence of different races was that, after the big flood, Noah's children spread to different parts of the earth. How we truly originated, and how physical differences appeared between us, were anyone's guess.

In 1871 biologist Charles Darwin published *The Descent of Man*, sweeping away these religious creation myths and framing the human species as having had one common ancestor many millennia ago, evolving slowly like all other life on earth. Studying humans across the world, their emotions and expressions, he wrote, 'It seems improbable to me in the highest degree that so much similarity, or rather identity of structure, could have been acquired by independent means.' We are too alike in our basic responses, our smiles and tears, our blushes. On this alone, Darwin might have settled the race debate. He demonstrated that we could only have evolved from shared origins, that human races didn't emerge separately.

On a personal level, this was important to him. Darwin's family included influential abolitionists, his grandfathers Erasmus Darwin and Josiah Wedgwood. He himself had seen the brutality of slavery first-hand on his travels. When naturalist

Louis Agassiz in the United States spoke about human races having separate origins, Darwin wrote disparagingly in a letter that this must have come as comfort to slaveholding Southerners.

But this wasn't the last word on the subject. Darwin still struggled when it came to race. Like Abraham Lincoln, who was born on the same day, he opposed slavery but was also ambivalent on the question of whether black Africans and Australians were strictly equal to white Europeans on the evolutionary scale. He left open the possibility that, even though we could all be traced back to a common ancestor, that we were the same kind, populations may have diverged since then, producing levels of difference. As British anthropologist Tim Ingold notes, Darwin saw gradations between the 'highest men of the highest races and the lowest savages'. He suggested, for example, that the 'children of savages' have a stronger tendency to protrude their lips when they sulk than European children, because they are closer to the 'primordial condition', similar to chimps. Gregory Radick, historian and philosopher of science at the University of Leeds, observes that Darwin, even though he made such a bold and original contribution to the idea of racial unity, also seemed to be unembarrassed by his belief in an evolutionary hierarchy. Men were above women, and white races were above others.

In combination with the politics of the day, this was devastating. Uncertainty around the biological facts left more than enough room for ideology to be mixed with real science, fabricating fresh racial myths. Some argued that brown and yellow races were a bit higher up than black, while whites were the most evolved, and by implication, the most civilised and the most human. What was seen to be the success of the white races became couched in the language of the 'survival of the fittest', with the implication that the most 'primitive' peoples, as they were described, would inevitably lose the struggle for survival as the human race evolved. Rather than seeing evolution acting to make a species better adapted to its particular environment, Tim Ingold argues that Darwin himself began to frame evolution as an 'imperialist doctrine of progress'.

'In bringing the rise of science and civilisation within the compass of the same evolutionary process that had made humans out of apes, and apes out of creatures lower in the scale, Darwin was forced to attribute what he saw as the ascendancy of reason to hereditary endowment,' writes Ingold. 'For the theory to work, there had to be significant differences in such endowment between "tribes" or "nations".' For hunter-gatherers to live so differently from city-dwellers, the logic goes, it must be that their brains had not yet progressed to the same stage of evolution.

Adding fuel to this bonfire of flawed thinking (after all, we know that the brains of hunter-gatherers are no different from those of anyone else) were Darwin's supporters, some of whom happened to be fervent racists. The English biologist Thomas Henry Huxley, known as 'Darwin's Bulldog', argued that not all humans were equal. In an 1865 essay on the emancipation of black slaves, he wrote that the average white was 'bigger brained', adding, 'The highest places in the hierarchy of civilisation will assuredly not be within the reach of our dusky cousins.' For Huxley, freeing slaves was a morally good thing for white men to do, but the raw facts of biology made the idea of equal rights – for women as well as for black people – little more than an 'illogical delusion'. In Germany, meanwhile, Darwin's loudest cheerleader was Ernst Haeckel, who taught zoology at the University of Jena from 1862, and was a proud nationalist. He liked to draw connections between black Africans and primates, seeing them as a kind of living 'missing link' in the evolutionary chain that connected apes to white Europeans.

Darwinism did nothing to inhibit racism. Instead, ideas about the existence of different races and their relative superiority were merely repackaged in new theories. Science, or the lack of it, managed only to legitimise racism, rather than quash it. Whatever real and reasonable questions might have been asked about human difference were always tainted by power and money.

*

I pick my way through a tall thicket of bamboo and find an intricate wooden pagoda.

Further still inside the sunlit Garden for Tropical Agriculture is a Tunisian house, coated in thick green moss. If their histories were unknown to me, I might find the buildings in this quiet maze beautiful. They are grand and otherworldly, ethereal relics of foreign places as imagined by another age. But of course, I'm acutely aware that each was also once a kind-of home to real people like me, pulled from their lives thousands of miles away for the entertainment of paying visitors. As a reminder, through the smashed window of a Moroccan castle, complete with battlements and blue tiles, I'm caught off-guard by a glaring red face that must have been painted by vandals.

However beautiful they are, these aren't homes at all. They're gilded cages.

It's hard to imagine what life would have been like on the inside of the human zoos, looking out. The people kept here weren't slaves. They were paid, similar to actors under contract, but expected to dance, act, and carry out their everyday routines in public view. Their lives were live entertainment. They were objects first and people second. Little effort was made to help them feel comfortable in their temporary homes, much less to acclimatise them. After all, the whole point of the spectacle was to underscore just how different they were, to imagine that even in a cold climate they would choose to walk around in as few clothes as they wore in a hot one, that their behaviour couldn't change no matter where they lived. Visitors were made to believe that the cultural differences were woven into their bodies like stripes on a zebra. 'When there was a birth, it meant a new show,' Gilles Boëtsch from the National Center for Scientific Research tells me. People would flock to see the baby.

Science had created a distance between the viewers and the viewed, the colonisers and the colonised, the powerful and the powerless. For those confronted with people from foreign lands in this way, bizarrely out of context, referenced in a book or transplanted to some fake village in Paris, it only helped reinforce the notion that we were not all quite the same. For the spectators peering into their homes, the performers in human zoos must have been curiosities not just because they looked and behaved differently, but because control of their lives belonged to others who didn't look like them. The ones outside the cage were clothed, civilised and respectable while those inside were semi-naked, barbaric and subjugated.

'People are more readily perceived as inferior by nature when they are already seen as oppressed,' write American scholars Karen Fields and Barbara Fields in their 2012 book *Racecraft*. They explain how a sense of inevitability gets attached to a social routine until it becomes seen as natural. The idea of race didn't make people treat other people as subhuman. They were already treated as subhuman before race was invoked. But once it was invoked, the subjugation took on a new force.

There was something about treating human difference as a science that gave it a peculiar quality. The observation of humans turned humans into strange beasts. While the unimpeachable impression of scientific objectivity was maintained, somehow the gold standard of beauty and intelligence always turned out to be the scientist himself. His own race was safe in his hands. German naturalist Johann Blumenbach, for instance, idealised the Caucasian race to which he belonged, but described Ethiopians as being 'bandy-legged'. If legs were different, there was never any question that Caucasians might be the unusual ones. The creatures caged in the human zoos were those who had failed to reach the ideal of white European physical and mental perfection.

The scientific distance created by believing that racial hierarchies existed in nature, this uneven balance of power, allowed human zoos to treat their performers as less than equals, making life for them fatally precarious. According to Boëtsch, many died from pneumonia or tuberculosis. Concerns were expressed in the press. There were always protests, as there had been about Saartjie Baartman, but they made little difference.

In another example around the same time as the Paris Exposition, a Congolese 'pygmy' named Ota Benga, who had been brought to the United States to be displayed at the St Louis World's Fair, was put in the Monkey House at Bronx Zoo in New York, without shoes. Visitors loved him. 'Some of them poked him in the ribs, others tripped him up, all laughed at him,' the New York Times reported. He was eventually rescued by African American ministers, who found him a place in an orphanage. Ten years later, in despair because he couldn't return home to the Congo, he borrowed a revolver and shot himself through the heart.

As I stand among the weeds and crumbling former homes of Paris's human zoo, it's difficult to avoid concluding that the reason anyone pursued the scientific idea of race was not so much to understand the differences in our bodies, but to try to justify why we lead such different lives. Why else? Why would something as superficial as skin colour or body shape matter otherwise? What the scientists really wanted to know was why some people are enslaved and others free, why some prosper while others are poor, and why some civilisations have thrived while others haven't. Imagining themselves to be looking object- ively at human variation, they sought answers in our bodies to questions that existed far outside them. Race science had sat, always, at the intersection of science and politics, of science and economics. Race wasn't just a tool for classifying physical dif- ference, it was a way of measuring human progress, of placing judgement on the capacities and rights of others.

3

Scientific Priestcraft

Deciding that races could be improved,
scientists looked for ways to improve their own

THE PAST IS BUILT of the things we choose to remember.

The Max Planck Society, with its headquarters in Munich, Germany, has an illustrious history. It has been the intellectual home of eighteen Nobel Prize winners, including the theoretical physicist Max Planck, after whom it's named. With an annual budget of 1.8 billion euros, its institutes employ more than 14,000 scientists, producing over 15,000 published scientific papers a year. By any standards, it's one of the most prestigious centres of science in the world. But in 1997 biologist Hubert Markl, then president of the Max Planck Society, made a decision that would threaten the reputation of his entire establishment. He wanted to scratch beneath its glorious history to reveal a secret that had been hidden for fifty years.

Before 1948, the Max Planck Society existed in a different incarnation, the Kaiser Wilhelm Society. Established in the German Empire in 1911, it was as important then as it is now, cementing Germany's place in modern scientific history. Even

Albert Einstein did some of his research at one of its institutes. But it was later, as the Nazis took power and began to act on their own scientific priorities, that things took a disturbing turn. We know that figures from within science and academia must have played a role in developing Adolf Hitler's ideology of racial hygiene, which argued that those of pure, 'Aryan' racial stock should be encouraged to breed, while others were gradually eliminated – an ideology that culminated in the Holocaust. It couldn't have been done without scientists, both to provide the theoretical framework for such an audacious experiment, and to carry out the job itself. On the practical side, there would have been those setting up concentration camps and gas chambers, as well as determining who should die. And then there were all the gruesome human experiments known to have been carried out on people who were eventually killed, plundering them for biological data.

There were rumours that staff from within the Kaiser Wilhelm Society had been involved, that they were maybe even party to murder and torture. In hindsight, they must have been. Under the regime, notes writer James Hawes, half the nation's doctors were Nazi party members. For a decade, German universities had taught racial theory.

But whatever went on was quietly forgotten after the Second World War. Although there was undoubtedly a story to be uncovered, it was thought wiser to leave it alone. By the Max Planck Society's own admission, it had a tradition of glossing over its ignominious past in favour of celebrating its greater scientific achievements. By the 1990s, however, there was too much pressure from the public to ignore that past any longer. And anyway, older members of staff who had been alive during the war – who might be affected by such revelations – had almost all died. The time had come. So Markl resolved to lift the lid, appointing an independent committee to investigate what German scientists at the Kaiser Wilhelm Society might have done during the war. It would be an investigation into the very darkest corners of race science. Younger researchers at the Max Planck Society justifiably worried whether the body of scientific work they had inherited might bear bloody stains.

They were right to worry. The past turned out to be dripping with blood. A few years after Markl launched the investigation, historians began publishing their findings, and they were devastating. Some had assumed that the Nazis were ignorant of or hostile towards science. Historical evidence proved this wasn't true. The Kaiser Wilhelm Society's scientists had willingly cooperated with the Nazi state, marrying academic interests and political expediency, helping to secure financial support and social standing for themselves. 'Such research not only literally built on the spoils of war, it also led scientists deep into the abysses of Nazi crimes,' wrote a reviewer. At least one prominent scientist helped draft and disseminate the legislation relating to racial ideology.

Those who weren't opportunistic were often complicit, displaying moral indifference when they could see inhumane or criminal acts happening right in front of them. When moves began in 1933 to expel Jewish scientists from the Kaiser Wilhelm Society (Einstein abandoned Germany that same year, leaving for a conference and wisely never returning), staff made little effort to stand in the way. At least two of its scientists and two other staff members ended up dying in concentration camps.

And then there were those who wholeheartedly supported the Nazis from the beginning. The work of Otmar von Verschuer, head of department at the Kaiser Wilhelm Institute of Anthropology, Human Heredity and Eugenics, makes for chilling reading. Until the war, von Verschuer was a widely respected academic, his research on twins as a means of understanding genetic inheritance funded for a few years by the Rockefeller Foundation in New York. He was once invited to speak at the Royal Society in London. But he was also, it transpired, an anti-Semite who openly praised Hitler and believed in a biological solution to what he saw as the Jewish threat to racial purity. According to American anthropologist Robert Wald Sussman, von Verschuer became one of the Nazis' race experts when it came to addressing the 'Jewish question', actively legitimising the regime's racial policies. One of his former students, the doctor Josef Mengele, went on to become infamous for his cruel experiments on twins and pregnant women at Auschwitz concentration camp. British writer Marek Kohn has documented in his 1995 book *The Race Gallery* that among the samples sent to von Verschuer from Auschwitz were 'pairs of eyes from twins . . . dissected after their murder . . . children's internal organs, corpses and the skeletons of murdered Jews'.

In 2001, the Max Planck Society at last accepted responsibility for historic crimes committed by its scientists. In its apology, the society admitted, 'Today it is safe to say that von Verschuer knew of the crimes being committed in Auschwitz and that he, together with some of his employees and colleagues, used them for his purposes.' Markl added in his speech, 'The Kaiser Wilhelm Society tolerated or even supported research among its ranks that cannot be justified on any ethical or moral grounds . . . I would like to apologise for the suffering of the victims of these crimes – the dead as well as the survivors – done in the name of science.'

This came too late for justice, of course. Those involved had died already. What was remarkable was that it had taken so long to root out the facts, to even find the will to do it. Scientists complicit with the regime had been skilled at covering their tracks, evidently. But maybe it was also easier for their colleagues to pretend that fellow scientists couldn't possibly have been active participants in murder and torture. Perhaps, they imagined, they were just bystanders, caught up in the mess while trying to get their work done.

The truth – that it is perfectly possible for prominent scientists to be racist, to murder, to abuse both people and knowledge – doesn't sit easily with the way we like to think about scientific research. We imagine that it's above politics, that it's a noble, rational and objective endeavour, untainted by feelings or prejudice. But if science is always so innocent, how is it that members of such a large and prestigious scientific organisation could have sold themselves to a murderous political regime as recently as the middle of the twentieth century?

The answer is simple: science is always shaped by the time and the place in which it is carried out. It ultimately sits at the mercy of the personal political beliefs of those carrying it out. In the case of some Nazi scientists, particular experiments may have been perfectly accurate and rigorous. They may even have produced good science, if goodness is measured in data and not human life. Other times, researchers didn't care about the truth or other people's lives, choosing instead to give the illusion of intellectual weight to a morally bankrupt ideology because it suited them.

Now, decades later, the horrors of the Second World War still have a warping effect on how we think about race science. Many of us choose to remember Nazi scientists like Otmar von Verschuer as some kind of uniquely evil exception, nothing like those who found themselves on the winning side of the war. The Holocaust and the twisted scientific rationale behind it are thought to belong to that time and place alone, purely the work of 'the bad guys'. But there was one question that went

unanswered after the investigations into the bloodstained his
tory of the Max Planck Society: Were scientists in the rest of the
world so blameless?

To file away what happened during the war as aberrant, as
something that could only have been done by the worst people
under the worst circumstances, ignores the bigger truth. This
was never a simple story of good versus evil. The well of scientific ideas from which Hitler and others in his regime drew
their plans for 'racial hygiene', leading ultimately to genocide,
didn't originate in Germany alone. They had been steadily supplied for more than a century by race scientists from all over
the world, supported by well-respected intellectuals, aristocrats,
political leaders and women and men of wealth.

Among the most influential of them all, as far as the Nazi
regime was concerned, was a pair of statisticians working at 50
Gower Street, Bloomsbury – not in Germany, but in the famous
old literary quarter of London.

*

'You have biologists who say there is no such thing as race, we
need to get over it, forget it,' Subhadra Das tells me in an angry
whisper. 'But then, if there is no such thing, why did you just say
"race"? Where did that idea come from?'

Das is a curator of the University College London Medical
and Science Collections, moonlighting occasionally as a stand-
up comedian. Her dark wit betrays a fury fed by the things she's
learned from her research. We're in the heart of Bloomsbury,
recognisable by its peaceful garden squares and smart Geor-
gian townhouses. Once a meeting point for artists and writers,
including Virginia Woolf, it is still home to a large slice of London's universities and colleges. Outside, busy Gower Street is
jam-packed with students heading for lectures, but where Das

and I are it's library quiet. We're seated at a small table inside the Petrie Museum, named for Sir Flinders Petrie, an Egyptologist who, before he died in 1942, used to collect heads from around the world to shore up his ideas of racial superiority and inferiority.

'Scientists are socialised human beings who live within society, and their ideas are social constructions,' she continues. She wants me to hear this, setting the scene before she begins unfolding the packets of objects in front of us, which she has pulled from the archive. Among the first is a black-and-white photograph of a well-dressed older man, his bushy eyebrows resting in a canopy over his eyes, long white sideburns trailing down to his collar. Underneath is his autograph: it is the biologist Francis Galton, born in 1822, a younger cousin of Charles Darwin. Galton, she tells me, is the father of eugenics. He coined the term in 1883 from the Greek prefix 'eu' for 'well' or 'good', to describe the idea of using social control to improve the health and intelligence of future generations.

Galton considered himself an expert on human difference, on the finer qualities that make a person better or worse. Not quite the genius that Darwin was, he certainly aspired to be. 'I find that talent is transmitted by inheritance in a very remarkable degree,' he had written in an essay titled 'Hereditary Character and Talent'. His idea drew on his cousin's theory of evolution by natural selection, that individuals in a population show a wide variety of characteristics, but those with the characteristics most suited to the environment will survive and breed, passing on those beneficial traits. Galton thought that a race of people could be more quickly improved if the most intelligent were encouraged to reproduce, while the stupidest weren't – the same way you might artificially breed a fatter cow or a redder apple. For him, this would speed up human evolution, driving the race closer to mental and physical per fection.

As an example, he drew on the fact that brilliant writers were often related to other brilliant writers. He noted that of 605 notable men who lived between 1453 and 1853, one in six were related. The ingredients for greatness must be heritable, he reasoned, choosing to overlook that being notable might also be a product of connections, privilege and wealth, which these men also had. 'If a twentieth part of the cost and pains were spent in measures for the improvement of the human race that is spent on the improvement of the breed of horses and cattle, what a galaxy of genius might we not create!' Galton dreamed of a 'utopia' of highly bred super people, and he made creating one his lifelong mission.

The first challenge would be to measure people's abilities, to build up a bank of data about who exactly were the most intelligent and who the least. In 1904 he convinced the University of London to set up the world's first Eugenics Record Office at 50 Gower Street, dedicated to measuring human differences, in the hope of understanding what kind of people Britain might want more of. University College London jumped at the chance, replying to his request within a week. After a short time the department became known as the Galton Laboratory for National Eugenics.

Eugenics is a word that's no longer used around here. Long after Galton's death, his laboratory was renamed the Department of Genetics, Evolution and Environment, housed in the Darwin Building. And this is where Subhadra Das steps in. Among the vast collection of objects she is responsible for at the university is Galton's archive, containing his personal photographs, equipment and papers, tracking the genesis and development of eugenics. She also looks after objects belonging to his close collaborator, mathematician Karl Pearson, who became the first professor of national eugenics in 1911 after Galton died. 'Pearson's greatest contribution, the thing that people remember him for, is founding the discipline of statistics. A lot of work on that was done with Galton. Galton, if you're going to bring his science down to anything in particular, is a statistician,' she tells me.

But before he settled down into science, Galton had been an explorer. He was lavishly funded by the estate of his father, who had made a fortune from supplying weapons that helped support the slave trade, and later from banking. An expedition in 1850 to Namibia, then known as Damaraland, earned Galton a medal from the Royal Geographical Society. Always proud of his appearance (there's a hand mirror and sewing kit among his possessions in the collection), he donned a white safari suit, becoming one of the first to cultivate what is now the classic image of the white European in Africa. 'If I say to you "African explorer", the picture that pops into your head? That's him,' Das tells me.

What was unusual about Galton was that travel failed to broaden his mind. His encounters with people in other countries didn't help him to see their common humanity. 'If anything, his racist assumptions were made stronger by his time in Africa.' As Galton told the Royal Society on his return, 'I saw enough of savage races to give me material to think about all the rest of my life.'

In London, racism combined in his scientific research with a passion for data. Galton was obsessed with measuring things, once using a sextant to size up an African woman's proportions from a distance. Another time, he came up with the mathematical formula for the perfect cup of tea. Through eugenics he saw a way of using what he thought he knew about human difference, shored up by Darwin's theories of natural selection, to systematically improve the quality of 'the British race'. 'Darwin said that humans are animals like any other animal. Galton said, well, if that's the case then we can breed them better,' Das explains. 'What he was concerned about was what he saw as the degeneration of the British race and how that could be prevented and improved.

'You have to call Galton a racist because the work that he did is fundamental in the story of scientific racism. So not only is he a racist, he is part of the way we invented racism, and the way that we think about it.'

*

Eugenics is a cold, calculated way of thinking about human life, reducing human beings to nothing but parts of the whole, either dragging down their race or pulling it up. It also assumes that almost all that we are is decided before we are born.

The origins of this idea – that everything is inherited, that it's in the genes – date back to the middle of the nineteenth century when Gregor Mendel, an Augustinian friar in Brno, Moravia, then part of the Austro-Hungarian Empire, became fascinated by plant hybrids. Working in the garden of his monastery, Mendel took seven strains of pea and bred them selectively until each one produced identical offspring every time. With these true-bred pea plants, he began to experiment, observing carefully to see what happened when different varieties were crossed. Nobody knew about genes at this point, and Mendel's paper on the topic published in 1866 went largely unnoticed within his lifetime. But his experimental finding that traits such as colour were being passed down the generations in certain patterns would form the linchpin of how geneticists in the following century thought about inheritance.

Once scientists understood that there were discrete packets of information in our cells that dictated how our bodies were built, and that we got these packets in roughly equal measure from each parent, the science of heredity finally took off. And it took almost no time for the political implications to be recognised. In 1905 the English biologist William Bateson, Mendel's principal populariser, predicted that it 'would soon provide power on a stupendous scale'.

Mendelism became a creed, an approach to thinking about human biology which suggested that it is largely set in motion as soon as an egg is fertilised, and that things then go on to work in fairly linear fashion. If you crossed one yellow-seeded pea plant with one green-seeded pea plant and you could predict which colours subsequent generations of pea plant would turn out to have, then it stood to reason that you might be able to predict how human children would look and behave based on the appearance and behaviour of their parents.

Through a narrow Mendelian lens, almost everything is planned by our genes. Environment counts for relatively little because we are at heart the products of chemical compounds mixing together. We are inevitable mixtures of our ancestors. Just as Bateson foresaw, this idea became the cornerstone of eugenics, the belief that better people could be bred by selecting better parents. 'Mendelism and determinism, the view that heredity is destiny, they go together,' says historian Gregory Radick, who has studied Mendel and his legacy.

But there was a problem with Mendel's pea plant research. At the beginning of the twentieth century Mendel's paper became the subject of ferocious debate, says Radick. 'Should the Mendelian view be the big generalisation around which you hang everything else? Or on the contrary, was it an interesting set of special cases?' When Mendel performed his experiments,

he deliberately bred his peas to be reliable in every gener
ation. Before he even began, he filtered out the aberrations, the
random mutants, the messy spread of continuous variation you
would normally see, so every generation bred as true as possible.
Peas were either green or yellow. This allowed him to see a clear
genetic signal through the noise, producing results that were far
more perfect than nature would have provided.

Raphael Weldon, born in 1860, a professor at the University
of Oxford with an interest in applying statistics to biology,
spotted this dilemma and began campaigning for scientists to
recognise the importance of environmental as well as genetic
backgrounds when thinking about inheritance. 'What really
bothered him about the emerging Mendelism was that it turned
its back on what he regarded as the last twenty years of evidence
from experimental embryology, whose message was that the
effects a tissue has on a body depend radically on what it's inter-
acting with, on what's around it,' explains Radick. Weldon's
message was that variation matters, and that it is profoundly
affected by context, be it neighbouring genes or the quality of
air a person breathes. Everything can influence the direction
of development, making nurture not some kind of afterthought
tacked onto nature, but something embedded deep down in our
bodies. 'Weldon was unusually sceptical.'

To prove his point, Weldon demonstrated how ordinary pea
breeders couldn't come up with the same perfectly uniform
peas as Mendel. Real peas are a multitude of colours between
yellow and green. In the same way that our eyes aren't simply
brown or blue or green, but a million different shades. Or that if
a woman has a 'gene for breast cancer', it doesn't mean she will
necessarily develop the disease. Or that a queen bee isn't born a
queen; she is just another worker bee until she eats enough royal
jelly. Between the gene and real life is not just the environment,

but also random possibility. Comparing Mendel's peas with the real world, then, is like comparing a soap opera with real life. There is truth in there, but reality is a lot more complicated. Genes aren't Lego bricks or simple instruction manuals; they are interactive. They are enmeshed in a network of other genes, their immediate surroundings and the wider world, this ever-changing network producing a unique individual.

Sadly for Weldon, the ferocious debate for the soul of genetics ended prematurely in 1906 when he died of pneumonia, aged just forty-six. His manuscript went unfinished and unpublished. With less resistance than before, Mendel's ideas were gradually incorporated into biology textbooks, becoming the bedrock of modern genetics. Although Weldon's ideas have since slowly been reincorporated into scientific thinking, there still remains a strain of genetic determinism in both the scientific and the public imagination. Harvard biologist Richard Lewontin has called it the 'Central Dogma of Molecular Genetics'. It is a belief that all that we are is set in stone in the womb.

In the early twentieth century, before the advent of modern genetics but with Mendel's findings prominent in their minds, Francis Galton's theories seemed to make good sense to many. They had a logical appeal that stretched across the political spectrum. We associate eugenics today with the fascists who per-petrated the Holocaust, but before the 1930s, many on the left saw it as socially progressive. Galton himself was certainly not considered a crank. He was a fellow of the Royal Society, and an anthropometric laboratory he set up in 1884 to catalogue people's measurements enjoyed support from the British Med-ical Association. Eugenics belonged firmly to establishment science, and amongst intellectuals, it wasn't just mainstream, it was fashionable.

The fly in the ointment was how to carry it out. Galton

observed that the poor seemed to be outbreeding the rich, and he saw the poor as poor for the simple reason that they were congenitally unfit. Responsible action was necessary to address the problem and ensure genetic progress. On the one hand, the rich needed to step up their baby-making game. On the other, society's dregs, particularly those described as mentally feeble, physically weak, and criminal types, needed convincing to have fewer children. Managing reproduction was the linchpin of eugenics, even attracting a fan in women's rights activist and birth control pioneer Marie Stopes. To support her first clinic, Stopes founded the Society for Constructive Birth Control and Racial Progress. Philosopher Bertrand Russell, too, suggested that the state might improve the health of the population by fining the 'wrong' type of people for giving birth.

Eugenics was more than a theory, it was a plan in search of policymakers. Winston Churchill, then First Lord of the Admiralty, was welcomed as vice-president at the first International Eugenics Congress, held at London's grand Hotel Cecil in 1912. Other vice-presidents included the Lord Mayor of London and the Lord Chief Justice. Delegates came from all over Europe, Australia and the United States, including Harvard and Johns Hopkins University. The US state of Indiana had already passed the world's first involuntary sterilisation law in 1907, informed by eugenicists who argued that criminality, mental problems and poverty were hereditary. More than thirty other states soon followed, with enthusiastic public backing. By 1910 a Eugenics Record Office was established at Cold Spring Harbor on Long Island in New York, with support from oil industry magnate John D. Rockefeller and later funding by the Carnegie Institution of Washington.

A news item in the journal *Science* announced that one of the purposes of the new office in New York would be 'the study of miscegenation in the United States', the mixing and intermarriage of different racial groups. Its board of scientific directors included Alexander Graham Bell, inventor of the telephone, and the economist Irving Fisher. The hardware behind at least one of America's most ambitious eugenics projects came from none other than IBM, the same company that went on to supply the Nazi regime in Germany with the technology it needed to transport millions of victims to the concentration camps.

In the first decades of the twentieth century, all over the world, eugenics began to be conflated with nineteenth-century ideas about race. In Japan, Meiji-period thinker and politician Katō Hiroyuki used Darwinism to make the point that there was a struggle for survival between different nations. In China in 1905, the revolutionary Wang Jingwei argued that a state whose members were of a single race was stronger than one comprising multiple races. Other politicians advocated sterilisation as a means of human selection, and racial intermarriage to produce children with whiter skins. Historian Yuehtsen Juliette Chung has noted that during this time, 'China seemed to accept passively the notion of race as the West understood it.'

In India, too, European notions of racial superiority were easily absorbed by some, partly because they mirrored the country's existing caste system – itself a kind of racial hierarchy – but also because Germany's Aryan myth placed the noble race as having once lived in their region. The ideological quest for the true 'Aryans' remains alive in India, and Adolf Hitler's *Mein Kampf* is a bestseller in Indian bookshops. Each nation utilised the idea of race in its own ways, marrying it with science if it could be of use. Eugenics, then, became just another tool in what were longstanding power dynamics.

By 1914 the word 'eugenics' was being used with such abandon that it had almost became synonymous with being healthy, complained American eugenics professor Roswell H. Johnson in the *American Journal of Sociology*. 'A school for sex education is called a school of eugenics. Even a milk and ice station has been similarly designated,' he grumbled.

*

In its early days, particularly for its mainstream supporters, eugenics focused on improving racial stock by weeding out those seen to be at the margins of society, the feeble-minded, insane and disabled. But as time wore on, the umbrella inevitably expanded. Karl Pearson, who succeeded Galton as the main force behind eugenics when he died in 1911 and shared his views on race, believed that since other races than his own were inferior, intermixing was also dangerous to the health of the population. By this logic, the very existence of those other races represented something of a threat. 'Pearson's argument is that if you have uncontrolled immigration the welfare of British people is at stake,' Subhadra Das tells me.

At the time, despite the mainstream popularity of eugenics, some did notice the slippery slope. This is one reason why, despite all the support it attracted from politicians and intellectuals and how popular it became in other countries, eugenics never managed to gain a firm toehold in Britain and was not implemented by the government. British psychiatrist Henry Maudsley argued that privilege and upbringing could surely more accurately explain why some people were successful and others weren't. He noted that many remarkable people had unremarkable relatives. Another vocal critic was biologist Alfred Russel Wallace, who had come from humble beginnings to become an important and well-loved researcher, credited with formulating evolutionary theory at the same time as Darwin. 'The world does not want the eugenicist to set it straight,' he warned. 'Give the people good conditions, improve their environment, and all will tend towards the highest type. Eugenics is simply the meddlesome interference of an arrogant, scientific priestcraft.'

But it's important to remember that history might well have gone another way. Das pulls out another object from the archive. It's a narrow tin box, resembling a cigarette case but twice as long. It was brought to London by Karl Pearson, but had been designed by Eugen Fischer, a German scientist who had been director of the Kaiser Wilhelm Institute of Anthropology, Human Heredity and Eugenics. The box still bears Fischer's name. Inside is a neat row of thirty locks of artificial hair, ranging in colour from blonde (numbers 19 and 20) and light brunette in the centre, to bright red hair at one end and black Afro hair (number 30) at the other. At first glance it looks innocuous, like a colour chart you might find at the hairdresser's. But the disturbing story behind it is betrayed by the order in which the hair samples are placed. The most desirable colours and textures have been placed in the middle and the least acceptable at the margins. This simple little gauge tells a story of pure horror.

'Fischer used this device in Namibia in 1908 to establish the relative whiteness of mixed-race people,' reveals Das. In what is now remembered as the first genocide of the twentieth century, in the four years preceding 1908, Germany killed tens of thousands of Namibians as they rebelled against colonial rule. According to some estimates, up to 3,000 skulls belonging to those of the Herero ethnic group were sent back to Berlin to be studied by race scientists. 'Namibia was the first place that the Germans built a concentration camp. Depending on where your hair fell on the scale was the difference between life and death.' Similar methods would be used again, of course, a few decades later. Fischer's work would also go on to inform the Nuremberg Laws of 1935, outlawing intermarriage between Jews, blacks and other Germans. He became a member of the Nazi party in 1940.

Das takes out another box that belonged to Pearson, this time containing rows of glass eyes in different colours, framed in aluminium eyelids so eerily real that I fear one of them might blink. They are prosthetics of the kind that would have been fitted in patients who had eyes missing. In the context of eugenics, though, they served another purpose. 'This object, I have seen its twin brother on display in an exhibition about race hygiene in Germany at the Berlin Museum of Medical History at the Charité. This device was appropriated by Nazi scientists and, again, used to judge or measure race, particularly in Jewish people,' Das explains. 'You'll find photographs of Nazi scientists measuring people's heads, measuring people's noses, matching their eye colour.'

The eye and hair colour charts reveal just how slippery the dogged mantras of rationality and objectivity can be when it comes to studying human difference. 'Any scientist who claims that they are not politicised, or that they are asking questions out of pure curiosity, they are lying to themselves,' she continues. 'The structure in itself is fundamentally, structurally racist, because it has always been taken at its face. Never going back and taking apart those underpinnings.' What does it matter if one person has black hair and brown eyes, and another has blonde hair and blue eyes? Why not compare heights or weights or some other variable? These particular features matter only because they have political meaning attached to them.

In the United States, arguably the most racially charged place in the world at the time, evolutionary theory and eugenics came along at just the moment when intellectual racists could deploy them to full effect. Immigration into the US from countries considered to be undesirable had been curbed by the 1882 Chinese Exclusion Act, the country's first major law restricting immigrants. Twelve years later, three Harvard College graduates formed the Immigration Restriction League, arguing in favour of a literacy requirement for those who wanted to come to the US. The group's secretary, Prescott Farnsworth Hall, used Darwin's ideas of natural selection to caution against allowing into the country 'undesirable' immigrants who weren't 'kindred in habits, institutions and traditions to the original colonists'. In a lengthy racist tract in *The Annals of the American Academy of Political and Social Science* in 1904, he added, 'The doctrine is that the fittest survive; fittest for what? The fittest *to survive in the particular environment in which the organisms are placed*' (his emphasis).

By 1907 the Bellingham riots would see hundreds of white men, themselves recent arrivals from Europe, attack Indian immigrants living in the city of Bellingham in the state of Washington, blaming their 'filthy and immodest habits'. Reportedly, seven hundred Indians had to flee. The local *Bellingham Herald* complained, 'The Hindu is not a good citizen. It would require centuries to assimilate him, and this country need not take the trouble.'

It was against this backdrop that a new ideologue emerged. In 1916 a wealthy American law graduate named Madison Grant published a book that took eugenics to another level. Grant was known as a conservationist (as one of the co-founders of Bronx Zoo in New York, he had lobbied to put Congolese man Ota Benga on display among the apes there in 1906) but he wasn't a scientist. He recognised, however, the power of the language of science. In *The Passing of the Great Race:*

or *The Racial Basis of European History*, he revived the legacy
of Count Arthur de Gobineau from the previous century, promoting the myth of Aryanism. Grant proposed that a blond,
blue-eyed Nordic 'Master Race' represented the Aryans' true-life
descendants.

Grant's racial hierarchy was geographically specific, consigning to inferiority everyone who wasn't northern European
– including Italians and Greeks, who at that time were considered an undesirable immigrant group in the United States.
He warned against racial intermixing in the belief that this
would damage white racial purity even further. As casually as
a biologist writing about plant hybrids, he wrote that a cross
between any member of a European race and a Jew is a Jew.

In Grant, wealth and racism formed a toxic combination.
Being descended from some of the first European colonists to
settle in America, he counted himself – of course – among the
descendants of Aryans, a noble race under threat. Openly in
favour of both slavery and segregation, he made every possible
effort to reduce immigration to the United States from anywhere but northern Europe. And he had powerful supporters,
including soon-to-be President Theodore Roosevelt. In 1909
Grant became vice-president of the Immigration Restriction
League, of which Roosevelt was a member. In 1921 Grant was
the treasurer at the Second International Eugenics Conference
in New York.

And yet it took only the slightest interrogation for both his
historical and scientific evidence to be exposed as dodgy and
self-serving. One reviewer raised an eyebrow at Grant's claim
that Italian artists Dante, Raphael, Titian, Michelangelo and
Leonardo da Vinci were of the Nordic type, and that – stretching the geographical parameters even further – so was Jesus. But
the views of experts didn't matter to Grant's readers. His fake

assertions were enough for those seeking some apparently intellectual support in their opposition to immigration.

Two parallel ideologies had by now become firmly intertwined in the minds of racists. First, the decades-old concept of the existence of a superior race. Second, informed by eugenics, the idea that unless checked, inferior races would outbreed superior ones. Human variation, before the eighteenth century a jelly-like set of loose generalisations, had now become a hard matter of progress and struggle. Grant's work was referenced by the Ku Klux Klan. It also became one of the inspirations behind the Immigration Act of 1924, which set quotas according to nationality aimed at decreasing immigration from southern and eastern Europe, including Italy, Greece and Poland, as well as effectively barring anyone from Asia.

His work even earned one lifelong fan in Germany. In a fawning letter to Grant about *The Passing of the Great Race*, Adolf Hitler wrote, 'The book is my Bible.'

*

It was all so long ago, we imagine that it's well and truly over now. We think of the horrors of the Holocaust and earlier genocides, of slavery and colonialism, of the many millions who were killed, of the twisted logic behind these actions, as belonging to another time. We imagine that the end of the Second World War spelled an abrupt end for race science. Eugenics is a dirty word. We're enlightened now. We're wiser.

But the story doesn't end quite so quickly. While they may have tempered their politics, race scientists didn't simply disappear after the war. Those who had built their work around eugenics and studying human difference, who staked their careers on it, simply found new avenues.

Take Otmar von Verschuer, who had plundered the tiny

bodies of Auschwitz victims for his studies of twins during the Holocaust: after being temporarily banned from teaching, in 1951 he became professor of human genetics at the University of Münster. Many scientists similarly changed tack, gently manoeuvring themselves out of eugenics into allied fields, such as genetics, that studied human difference in less controversial and more rigorous ways. Many stopped using the word 'race' altogether. Science learned at least one lesson, recognising that if human variation was to be studied, it had to try and at least appear to stay away from politics.

But the shift didn't happen abruptly. The Eugenics Record Office on Gower Street in London survived all the way through the war. There is still a Galton Professor of Genetics at University College London, funded by money Francis Galton left behind. What was the Eugenics Society became the Galton Institute in 1989. In 2016, the institute established the Artemis Trust, which according to its own promotional leaflet, handed to me at a conference, distributes grants of up to £15,000, partly with the aim of assisting in the provision of fertility control, and particularly to those from 'poorer communities'.

Subhadra Das tells me that a woman came to see her recently whose mother had worked in the Galton laboratory in the 1950s. Her job had been to study redheadedness in Wales. It was not until the 1960s that the word 'eugenics' could no longer be heard in these corridors. What helped kill it in the end wasn't just the war but also the fact that new research showed it probably wouldn't work. The way we inherit traits from our parents turns out to be more complicated than Galton imagined. There is no guarantee that two beautiful and brilliant parents will produce brilliant and beautiful kids. Genetics is more of a game of chance. The science of inheritance, once it was better understood, didn't support the idea that humans could breed themselves to perfection, whatever perfection meant. Complex psychological traits such as intelligence are not controlled by a mere few genes, and are also heavily influenced by environment and upbringing.

Yet it was decades before eugenics policies introduced to other parts of the world were abandoned. Only in 1974 did the American state of Indiana repeal legislation that had made it legal to sterilise those it considered undesirable. Investigations by reporter Corey Johnson in 2013 uncovered that doctors working for the California Department of Corrections and Rehabilitation had continued the practice, sterilising as many as 150 women inmates between 2006 and 2010, possibly by coercing them into having the procedure. In Japan, a Eugenic Protection Law introduced in 1948 to sterilise those with mental illness and physical disabilities and prevent the birth of 'inferior' offspring was repealed only in 1996. Victims of the legislation are still pushing for justice.

The process of self-examination, of experiencing regret and showing remorse – the kind attempted by the Max Planck Institute in 2001 – is slow. And it has been particularly slow in the places that found themselves on the winning side in the Second World War. In the decades after the war, scientists in Britain and the United States airbrushed away their pivotal role in race science and eugenics. They quietly moved into other fields, silently renamed their university departments, consigning to the past that dark chapter. History was rewritten by the victors.

According to Gavin Schaffer, a professor of British history at the University of Birmingham and author of *Racial Science and British Society, 1930–62*, 'It was much easier to point the finger at the horrible Nazis, and the same went for the scientists. This absence of introspection was rooted in the ability to point fingers at other people for being responsible for the perversion of science.'

The post-war narrative of good triumphing over evil glossed over the messier truth that, in fact, everyone should have had a finger pointing at themselves. Without ever really looking back to the past and asking how and where the idea of race had been constructed in the first place, why it had been relentlessly abused, without questioning the motives of scientists like Francis Galton, Karl Pearson and countless others, in this glaring 'absence of introspection', old ideas of race could never completely disappear. Even long after the war, scientific fascination with human variation remained tainted by a lingering belief that there might be something deeper about racial difference, that perhaps some races really are better than others.

Yes, there was good science that emerged out of the ashes. Biology did attempt to reform itself, to cast away the mistakes of the past and do a more precise and accurate job of understanding human variation. But at the same time, while the world around them changed, a few of the hardened old-school race scientists could still be found knocking about. 'Racist science continues, it just becomes more marginal,' Gavin Schaffer tells me. 'But there's no doubt that it does continue.'

4

Inside the Fold

*After the war, intellectual racists
forged new networks*

THE SECOND WORLD WAR marked an unlucky turning point
in the life of scientist Reginald Ruggles Gates.

Born in 1882, Gates was one of those well-to-do, gentle-
manly race scientists who had been the norm in the nineteenth
century. He was a colonial type, who believed that other races
belonged to different human species, as well as a eugenicist
and supporter of segregation in the United States. One particular obsession was what he saw as the danger of different races
mixing with each other and having children. In his studies of
mixed-race people, he kept skin colour charts that looked like
paint swatches. Gates would almost certainly be considered a
racist by modern standards, but at the time his views weren't
uncommon. They certainly didn't get in the way of his career
or his standing in society. He was successful and well respected.

To get a sense of who Gates was I've come to the Maughan
Library at King's College London, to which his archive was
bequeathed after he died in 1962. Here, in a vast nineteenth-

century Gothic Revival building off Chancery Lane that was once Britain's Public Record Office, I leaf through his personal papers, slowly building a portrait. Sepia photographs show him to be smartly dressed, sporting a neatly clipped moustache. Gates had grown up in a wealthy family with thousands of acres of land across Nova Scotia in Canada before moving to Britain, where he was briefly married to Marie Stopes, a fellow member of the Eugenics Society. He enjoyed a career as a plant geneticist, becoming professor of botany at King's College in 1921, and later a fellow of the Royal Society.

He seems to have had a passion for travel, too, for understanding human difference across the world, his collection of scientific papers spanning almost every continent. In later life, he visited Cuba and Mexico to study 'mixed race' people, Japan and Australia to observe indigenous communities, and made a number of trips to India, a country that became a particular source of fascination. Browsing his personal collection of scientific papers in the library, I'm startled to discover there's even one on the blood groups of the Sainis in parts of Punjab – a study that may well have included relatives from my father's branch of the family.

Despite his racist beliefs, before the war Gates was riding high. Afterwards, though, it all changed. To his confusion and disappointment, he found himself left out in the cold by an establishment that had once welcomed him. His papers were rejected by scientific journals more often than they had ever been. And the reason was simple: deeply shaken by the genocidal use of eugenics by the Nazis in Germany, the world was turning its back on research that resembled their theory of racial hygiene. The enthusiasm for studying race that had once been almost fashionable was on its way out in scientific circles. Researchers who weren't wise enough to get with the new programme, who chose instead to cling to their unpalatable politics, as Gates did, found themselves flung from the warm centre of academic life to its chillier margins.

Yet he couldn't fathom it. 'What interests me about it is his incredulity,' I'm told by historian Gavin Schaffer. 'He seemed genuinely surprised.' In a sense, Gates was a man caught out by time. Francis Galton died in 1911 and Karl Pearson in 1936, before they could witness race science reach its most brutal peak. But others, such as Gates, lived long enough to see the political mood change – and then suffer the consequences when their own politics didn't.

At every opportunity, Gates refused to budge from his belief in racial superiority and inferiority. Wherever he found himself professionally hindered, he imagined himself to be the victim of some Jewish plot to derail his work. Schaffer recounts one especially bad experience in 1948 when Gates was working briefly at Howard University, the historically black college in Washington, DC. 'A petition was got up to remove him because of allegations that he was a racist – which he *was*. But he was stunned by that,' Schaffer says. 'He articulated his understanding of that as a manifestation of an international Jewish conspiracy, as opposed to just understanding that, in a historically black university, the kind of work that he did and the kind of things that he said were always going to be challenged.' Even when he agreed to leave Howard, Gates grumbled in private that only a few 'ignorant Negroes' were fit to be in a university at all.

Gates could never accept that the world was moving on, leaving those like him behind.

*

When it came to how the world thought about race, a wider political shift was underway. It was most clearly signposted in

1949 when more than a hundred scientists, anthropologists, diplomats and international policymakers met in Paris under the umbrella of the United Nations Educational, Scientific and Cultural Organisation, UNESCO, to redefine race. British-born American writer and anthropologist Ashley Montagu led the charge against scientific racism and its horrific legacy, taking his cue from a wave of social scientists who had already long argued that history, culture and environment were really behind what people thought of as racial difference.

'The word race is itself racist,' Montagu wrote in his influential 1942 book *Man's Most Dangerous Myth: The Fallacy of Race*. Both intellectually and culturally ahead of the curve, he explained in *American Anthropologist*, 'What a "race" is no one exactly seems to know, but everyone is most anxious to tell . . . The common definition . . . is based upon an arbitrary and superficial selection of external characters.' As anthropologists and geneticists were learning, individual variation within population groups, overlapping with other population groups, turned out to be so enormous that the boundaries of race made less and less sense. This was one reason why nobody had ever been able to agree on exactly how many races there were. Three, or four, or five, or several, there was never a consensus. The concept of race was as slippery as jelly, defying any effort to pin it down. In the end, academics had to concede that it probably wasn't an accurate or reliable way to think about human variation.

Montagu emphasised the likelihood that humans were genetically almost identical, and that in any case, our ancestral roots were certainly the same. Other anthropologists who had studied human diversity had already suggested that differences between humans were not only marginal, but also sat on continua, each so-called 'race' blurring into the next. What really made people and nations seem different was culture and language.

It was on the back of work like this that UNESCO, in July 1950, released its first statement on race, stressing unity between humans in a concerted effort to eradicate what it saw as the outcome of a 'fundamentally anti-rational system of thought'. It was meant to be the last word on the subject, to flush away racism once and for all. 'Scientists have reached general agreement in recognizing that mankind is one: that all men belong to the same species, *Homo sapiens*.'

The next few decades would be crucial to dismantling the idea that race was biologically real, and proving Montagu right. In 1972 a landmark paper exploring the true breadth of human biological diversity would appear in the annual edition of *Evolutionary Biology*. It was written by geneticist Richard Lewontin, who later became a professor at Harvard University. Dividing the planet up into seven human groups, based roughly on old-fashioned racial categories, Lewontin investigated just how much genetic diversity there was within each population, compared with the genetic diversity between them. What he found was that there was far more variation among people of the same 'race' than between the supposed races, concluding that around 85 per cent of all the genetic diversity we see sits within local populations, and 8 per cent more if you widen the net to continental populations. In total, around 90 per cent of the variation lies roughly within the old racial categories, not between them. There has been at least one critique of Lewontin's statistical method since then, but geneticists today overwhelmingly agree that although they may be able to use genomic data to roughly categorise people by the continent their ancestors came from (something we can often do equally well by sight), by far the biggest chunk of human genetic difference indeed lies within populations.

Lewontin's findings have been reinforced over time. An influential 2002 study published in *Science* by a team of American scientists, led by geneticist Noah Rosenberg, then at the University of Southern California, took genetic data from just over a thousand people around the world and showed that as much as 95 per cent of variation sits within the major population groups. Statistically this means that, although I look nothing like the white British woman who lives next door to me in my apartment building, it's perfectly possible for me to have more in common genetically with her than with my Indian-born neighbour who lives downstairs. Being of the same 'race' doesn't necessarily mean we are genetically more similar. In the long run, then, Ashley Montagu's position on race has been vindicated.

Mark Jobling, a respected professor of genetics at the University of Leicester, tells me that if there were a global catastrophe and all life were wiped out save just, say, Peruvians, 85 per cent of human genetic diversity would be safely retained. 'That just reflects the fact that we are a young species,' explains Jobling. Humans are relatively new, and being so new, we're still closely related to one another.

The greatest genetic diversity within *Homo sapiens* is found inside Africa, because this continent contains the oldest human communities. When some of our ancestors began to migrate into the rest of the world sometime between 50,000 and 100,000 years ago, the groups that moved were genetically less diverse than the ones left behind for the simple reason that they were made up of fewer people. The human variation we see across regions today is partly the result of this 'founder effect'.

Of course, groups of people have average physical differences, as a result of their biological and environmental histories. It has been estimated that 10,000 generations separate every single one of us from the original band of people in what is now Africa, but we vary in small ways because of the characteristics our ancestors happened to take with them as they migrated. As the small, sampled populations spread, bred and adapted to their local environments, they began to look more different from the relatives they left behind generations earlier and more like each other. And as small members of these groups themselves left for new territory, they would be slightly genetically different again because of a serial founder effect.

All this didn't happen in big clumps or clusters, but was rather more of a mesh, as people mated with those they encountered on the way, sometimes travelling further away and sometimes moving back. If everyone in the world had their genomes sequenced, adds Jobling, you wouldn't find hard borders between them, but gradients, with each small community blending into the next, the way hills blend into valleys. The racial categories we are used to seeing on census forms don't map onto the true picture of human variation.

The aim of the original 1950 UNESCO statement wasn't just to set out the science in a clear way, but to change the culture, to make people think differently about this idea they had lived with for so long, that had done incalculable damage to millions of lives. The statement emphasised that what we see as race is likely to be only a superficial variation on the same theme. Most of the visible variation is cultural. It tipped cold water over entrenched racial stereotypes, adding that there was no proof that groups of people differed in their innate mental characteristics, including intelligence and temperament.

This marked a crucial moment in history, a bold universal

attempt to reverse the deep-seated damage done by racism – and perpetuated by science – for at least two centuries. To some extent, it worked. Whether we realised it or not, all of us thought about race differently after that. Racism was no longer accept- able. Scientists and anthropologists by and large got behind UNESCO, and their work in the coming decades would in the main reflect that.

But this wasn't the end of it. And that's because not everyone was on board.

*

Despite the changing public mood about race, some re- searchers just couldn't bring themselves to ditch a body of work they had been cultivating for decades. Many didn't agree with UNESCO's claim that biology supported the idea of a universal brotherhood. A few couldn't accept that there were no mental differences between racial groups. And they weren't necessarily all racists. Some of them were respectable, eminent scientists at universities including Oxford and Cambridge, who simply wanted the statement to be revised with more scientific pre- cision and qualification. But one of the most passionate voices of all belonged to Reginald Ruggles Gates.

'What Gates called for, time and again, was the objective continuing study of race . . . because he thought that his pos- ition was grounded in *true* science,' explains Gavin Schaffer. He and others felt that UNESCO was stepping outside the bounds of what biology could actually claim, that it was ignor- ing facts in favour of liberal, anti-racist politics. 'The biologists who countered it, what they wanted was the continuation of their own expertise, which they had asserted over twenty, thirty years. They agreed the Nazi state was completely wrong in the way it had used race, and that other political actors had been completely wrong, but they felt that the study of race would profit from further work. I think, to them in that period, they felt they wanted work on race to continue, and they felt that other people wanted it to stop.'

The pressure worked, at least to some extent. In 1951 UNESCO gathered a team of experts to publish a new statement, tempering its language to account for the lack of consensus around the biological facts. The changes were subtle but revealing. For instance, instead of saying that scientists had 'reached general agreement' that we were one human species, the revised statement was gently altered to say that scientists were 'generally agreed . . .' In short, it had to make clear that not every expert could accept even the most basic fact that we all belonged to the same species.

Despite concessions like these, Gates failed to keep race science alive in the same way as before. By now, it had been all but lifted out of the laboratory. The academic study of race no longer sat within the realms of biology. Whether all biologists liked it or not, by the second half of the twentieth century race belonged to the social sciences, to the study of culture and history. It was understood to be a social and political construction, not borne out by biology. Old-fashioned race researchers and eugenicists had to move on or be sidelined.

At that point, Schaffer explains, 'The biologists just go into themselves a bit. They go back to their work, they go back to their labs.' They moved into newer fields, such as genetics, evolutionary biology and psychology. For those who still wanted to study human variation, their research moved away from skin colour and hair texture and went to the molecular level. 'As long as you weren't hell-bent on the kind of politics that were going to call your position into threat, yes, why not.' The older, somewhat cruder and more controversial ways of studying human difference, using anatomy and twin studies, were treated with suspicion. By the 1950s the word 'race' was so unfashionable in scientific circles that it was barely used any more.

Historian Veronika Lipphardt at University College Freiburg, Germany, has noted that the 1950s saw new institutes dedicated to the study of human variation open across the world. There was one in Bombay, another at Columbia University in New York, and one at the Federal University of Paraná in Brazil. A politically correct scientific terminology emerged. Researchers began referring to 'populations', and occasionally 'ethnic groups'. But the departure from the old race science wasn't quite as complete as it might have been. Although the parameters of research had changed, racial categories were still alive in people's minds. They were still active in everyday life, playing out in the politics and racism of the real world. For scientists to suddenly stop thinking about humans in racial terms was impossible so long as everyone out there still thought about themselves and others that way. So they couldn't help but look for racial difference, to subconsciously force this way of thinking into their work.

One example is blood type. When genetics became the preferred way to talk about human variation, hard hereditary variables like blood type came under the spotlight. Categorising by blood sounded more mathematical, less wishy-washy, than talking about skin colour or hair texture. And in the process, blood became an obsession. It was already well known that the proportions of people with different blood types varied from population to population, because of a phenomenon known as genetic drift. In prehistory, as small founding communities of people migrated across the world, they took their own narrow subset of blood types with them. This is equivalent to your cousin, say, leaving home to set up a colony of her own. She may be closely related to you, but have a different blood type. As these communities grew bigger, their particular blood types became the common ones. For example my own, B+, is shared by around a third of people in India where my family are from, and less than a tenth of people in the United Kingdom, where I live. By studying which population groups have which blood types, researchers found they could open a window into how closely or distantly related these groups might be to each other.

In the post-war period, the distribution of blood types became a hot topic in anthropology journals. In the 1960s the World Health Organisation launched its own effort to document groups of people around the world, collecting data on skin pigmentation and hair form, but also on blood type, colour blindness and other genetic markers. When the blood of the Sainis of Punjab was collected and tested in 1961 by a pair of anthropologists at the University of Delhi, before the results were passed to Reginald Ruggles Gates in England, it was part of these bigger efforts. Similar tests were carried out in other Indian groups and castes. Ultimately, thousands of people from different communities would have been gauged in the same way. And the same happened all over the world. At least some of these scientists were searching for proof that race was genetically tangible, that evidence for deep racial differences could be found at the molecular level.

Gates was one of those. He just couldn't let go of his belief in the old, hard, biologically rooted racial gaps. And he would never change. His final work, entitled *The Emergence of Racial Genetics*, published posthumously in 1963, attempted to place the new genetics in the old framework of race. One reason he pressed on with his commitment to races as meaningful categories, argues Schaffer, is that he sincerely believed his own research was objective, and those challenging him were the ones driven by ideology. He saw himself as the bearer of truth, held back by an anti-science political agenda that was mistakenly trying to impose racial equality on the world. He believed it even while he was receiving funding from segregationists in the United States.

Schaffer reminds me that it's important to understand the psychology behind this. When Gates complained that race research was being politicised, he was right. After all, following the Second World War and the bald brutality of the Holocaust, it would have been bizarre for any discussion of race not to be affected by politics. But what Gates failed to accept was that he was similarly affected by his own politics. 'People like him must also position themselves within that model. People who defend race historically have also done so for political reasons,' he explains. 'The science never becomes separated from political discourse.'

Gates wasn't a pseudoscientist, even if he was a bit of a crank. But his failure to gain professional recognition after the war wasn't just about his beliefs. Editors of some of the scientific journals rejecting his work warned him that his methods were becoming sloppy, relying on subjective interpretation rather than rigorous, intensive study. This slapdash approach may have been acceptable the previous century, but it didn't cut it in the world of modern science. By the end, both for his abhorrent views and his weak research, Gates had few supporters left in the scientific community. Reportedly, when his death was announced in 1962 at a meeting of American anthropologists, there were cheers.

That said, history doesn't move in a straight line. Ideas, even the worst ones, can go out of fashion in one century and come back in another. Those who imagined that the end of the war marked the abrupt death of race science were sadly mistaken.

*

In the final years before his death, when few scientific journals would touch his work, Gates decided to take matters into his own hands. If they wouldn't publish him, he would publish himself.

He and a handful of likeminded researchers, some on the very darkest margins of science – including former Nazi scientist Otmar von Verschuer (who died soon after Gates in a car accident, in 1969) and British eugenicist Roger Pearson (the last of the group still alive today, aged ninety, although he declined to give me an interview for reasons of ill health) – set up a journal of their own. Their aims were simple: to challenge what they saw as a politically correct, left-wing conspiracy around race and bring back some scientific objectivity. They named their brave new enterprise the *Mankind Quarterly*.

The founders of the journal regarded themselves as 'the defenders of the truth', says Schaffer; they even compared it to the gospels. But it would have been immediately clear to anyone who read it that the *Mankind Quarterly* was less impartial than it claimed to be. By 1960, when it launched, South Africa was in the midst of apartheid, the US civil rights movement was gaining momentum, and European colonies in Asia and Africa were winning independence. Race was high on the agenda everywhere, and the moral failures of the past were slowly being redressed. For racists who didn't welcome this shifting tide, now was the moment to assert their position. And the *Mankind Quarterly* was happy to oblige. It waded deep into the politics of the time, using science – even if only in a loose way – as its weapon of choice.

Recruiting truly respectable scientists to the cause was a challenge. But not an impossible one. The earliest editions included articles by Henry Garrett, a former president of the American Psychological Association and head of Columbia University's psychology department. Garrett was then one of the most powerful and eminent voices against desegregation in the United States. Most notably, in 1954, he had testified to stop the integration of black and white schools in the state of Virginia, when a trial known as Davis v. County School Board of Prince Edward County went up to the Supreme Court. Hundreds of students at an underfunded all-black school, with no gym or cafeteria, who were sometimes forced to study in an old school bus, fought against separate schools on the grounds that they were being disadvantaged because of their colour. The judge ruled against them. Under national pressure, the case was later combined as one of five brought before the Supreme Court in 1954 in the landmark action Brown v. Board of Education, which finally declared that separate schools for black and white students were unconstitutional.

Writing in the *Mankind Quarterly* in 1960, Garrett wasn't rolling back from his defeat, he was doubling down. Regardless of what the law said, for him, people of different races mixing with each other spelled certain disaster. 'The weak, disease-ridden population of modern Egypt offers dramatic evidence of the evil effects of a hybridization which has gone on for 5,000 years. In Brazil, coastal Bahia with its negroid mixtures is primitive and backward as compared with the relatively advanced civilization of white southern Brazil,' he wrote.

In another article for the journal in 1961, Garrett laid into academics, politicians and social reformers who didn't accept the 'common-sense' judgement that 'the Negro' was 'less intelligent and more indolent than the white'. Like so many scientific racists before him, Garrett argued that this was a matter of civilisational superiority. He claimed that Africans had never produced anything of great value. Could any African Negro, he charged, 'compare with the best of the European whites . . . for example, with Aristotle, Cicero, Thomas Acquinas [sic], Galileo, Voltaire, Goethe, Shakespeare or Newton?'

The twisting of facts to suit an ideological viewpoint would become a regular feature of the *Mankind Quarterly*. An especially cold-blooded article in 1966, this time by one of the editors, argued that Aboriginal Australians and Native Americans had been all but wiped out by European colonisers not out of greed or cruelty, but because it was a natural outcome of biology. 'If the conquered are markedly inferior to the conquerors . . . they will always remain an outcaste element at the bottom of the social structure.' Inter-racial conflict, the writer argued, was the product of natural selection, the fittest fighting to survive. Drawing parallels with the American civil rights movement, he added that it was virtually self-evident that racial integration would never work.

Articles like these didn't go unnoticed in the scientific community. Almost as soon as the *Mankind Quarterly* appeared, disgusted anthropologists sent in letters of complaint, accusing the journal of trying to make scientific racism respectable again. Slovene anthropologist Božo Škerlj, who had mistakenly joined the *Mankind Quarterly*'s advisory board only to be appalled by its 'ostensibly racialist editorial policy', entered into a public spat with the editors. Škerlj was particularly insulted by Gates's accusation that his mental outlook – and presumably his objectivity – was affected by the fact that he had been imprisoned in Dachau concentration camp during the war. Gates noted, revealingly, that he would never have considered Škerlj for the position in the first place had he known about his internment.

In *Science*, one of the world's leading journals, a reviewer called on scientists to take action. But it made no difference. The reason the *Mankind Quarterly* had been created at all was the lack of scientific approval for the kinds of ideas its founders wanted to publish. The editors didn't want or care about approval, they just needed a platform.

The other, deeper secret behind the *Mankind Quarterly* was that it had legs of its own. Support came indirectly from a reclusive, multimillionaire American textile heir with a vested political interest in what it was publishing. Wickliffe Draper was a diehard segregationist descended from a commanding officer in the Confederate Army on one side and the largest slaveholder in the state of Kentucky on the other. His family roots in America dated back to 1648, with enormous wealth and property steadily amassed over the centuries. In his 2002 book *The Funding of Scientific Racism*, William Tucker, emeritus professor of psychology at Rutgers University, details Draper's upbringing – one so privileged that a relatively weak academic record wasn't enough to stop him getting into Harvard University. There, as Tucker writes, he would have been exposed to those at the forefront of the American eugenics movement in the early twentieth century. An intellectual racist looking for ways to spend his inheritance, Draper would find in the *Mankind Quarterly* the perfect vehicle for his racist views.

In March 1937, Draper incorporated the Pioneer Fund, a private foundation aimed at disseminating information on human heredity and eugenics, providing race scientists who couldn't find backing anywhere else with the cash they needed to carry on. American anthropologist Robert Wald Sussman explains in his 2014 book *The Myth of Race* that 'Draper wanted to recruit scientific authorities with academic credentials and scholarly records who believed in the necessity of racial purity and that integration posed a threat to civilisation.' In short, he was trying to build a scholarly argument to defend segregation. During the war, his money also helped distribute a Nazi propaganda film about eugenics to US schools and churches. But it was after the war that the fund really came into its own. In 1959 he set up what he called the 'International Association for the Advancement of Ethnology and Eugenics', to produce and publish documents on race. The association's aim was to promote and distribute the *Mankind Quarterly*, helping turn it into one of the most important channels for race research in the world. William Tucker describes the association's original American directors as 'probably the most significant coterie of fascist intellectuals in the postwar United States and perhaps in the entire history of the country'.

The Pioneer's funding priority from the beginning was to back distinguished scientists, the more well known the better, along with racist ideologues. 'Grants to the former were intended to provide a façade of intellectual respectability for the latter, as well as results that could be used to justify their policies,' Tucker tells me. Cash gifts were routinely made to scientists who echoed Draper's political sentiments, while thousands of copies of the *Mankind Quarterly* containing their work were sent out to a list of American political conservatives. The science and the politics operated hand in glove.

Unsurprisingly then, according to Tucker, the journal made no concessions to political correctness. 'This was going to be a publication frankly written *by* racists *for* racists,' he writes. The target audience didn't appear to be the academic community at all, but racist movements searching for evidence that their prejudices might be rooted in scientific fact. 'Nothing seemed too bizarre or too repugnant to receive the *Mankind Quarterly*'s stamp of approval.' Among the lengthiest articles it published was one entitled "The New Fanatics', slamming American intellectuals who used their authority to support equal rights for black people. Sussman has noted that the book review section was in essence a bulletin board for publications that had anything to do with eugenics, lavishing praise on new publications that were neo-Nazi, anti-Semitic or anti-black.

To those who read the journal, its editors' intentions would have been clear. Many of the mainstream scientists who did bother to read it saw straight through it. A scholarly review of the first three editions by the late British anthropologist Geoffrey Ainsworth Harrison, a president of the Royal Anthropological Institute, was scathing. He complained that one of the editors hadn't grasped the concepts of modern genetics, despite writing about them at length. He dismissed Henry Garrett's work, too, as full of inconsistencies. Although Harrison didn't rubbish the value of studying human variation altogether, he didn't see what the Mankind Quarterly did as academically useful. 'Few of the contributions have any merit whatsoever, and many are no more than incompetent attempts to rationalise irrational opinions . . . it is earnestly hoped that The Mankind Quarterly will succumb before it can further discredit anthropology and do more damage to mankind,' he concluded.

But that didn't happen. In fact, it kept going for many more decades, publishing scientists and not-quite-scientists at the margins of their fields, many of them bankrolled by Wickliffe Draper's Pioneer Fund. The fund stuck to its aims even after Draper died in 1972. In his will, he left $50,000 to Henry Garrett alone. Despite all the criticisms it faced when it was first published, despite the widespread expectation that it wouldn't last, the Mankind Quarterly never succumbed. Indeed, if you want to read it, it's still around today.

*

When I contact German biochemist Gerhard Meisenberg, the current editor-in-chief of the *Mankind Quarterly*, I don't expect to hear back from him. After all, this is a journal considered so inflammatory that the security filter of my home broadband provider won't even allow me to look at its website without changing my settings. So I'm surprised not only to hear from Meisenberg immediately, but that he seems perfectly happy to tell me whatever I want to know.

He advises me that he became editor of the *Mankind Quarterly* only within the last few years, and that his job is to 'start the seemingly hopeless task to salvage this run-down journal', betraying the possibility that he sees renewed interest for the ideas it publishes. At the same time, he warns me that he can only communicate with me through email – not because he considers the work he does to be inflammatory, but because it's tricky to reach him by any other means right now. Since 1984 he's worked at the Ross University School of Medicine, a for-profit private college based in Dominica, but he and his students recently found themselves kicked off the island by a hurricane. As a result, he's teaching from a rented cruise ship elsewhere in the Caribbean.

I can't call or see him, he says, but we can write to each other. What follows is a long and candid exchange.

'I can tell you about how modern races evolved,' he tells me in his first message. It's clear from the outset that he believes he has an understanding of the subject not shared by mainstream scientists. The reason he is so happy to talk is because I'm giving him the opportunity to enlighten me. 'One hang-up for academic definitions of race is that academics like precise definitions and precise boundaries between categories. They use only their left brain hemisphere. The right one is atrophied. For them, when categories into which they slice the world lack clear boundaries, they seem to assume that the categories are invalid.'

On school performance in the United States, he states, 'Jews tend to do very well, Chinese and Japanese pretty well, and Blacks and Hispanics not so well. The differences are small, but the most parsimonious explanation is that much and perhaps most of this is caused by genes.' There is no scientific evidence for this, it's just speculation. Nobody has ever found any genes linking ethnicity or race to school results. Like Henry Garrett half a century earlier, Meisenberg chooses to skip over the social, historical and economic aspects of racial inequality. Rather, he believes that scientific evidence that doesn't yet exist will explain the gaps eventually. He takes it as given that the answer must be biological.

'Of course we can use molecular genetics to figure out, for example, in what way intelligence-related genetic variants vary among different racial groups,' he suggests. 'This will answer the question about race differences in intelligence once and for all. Good riddance of a stupid debate!'

He continues in this vein, the occasional sensible observation punctuated by more bizarre ones. At one point he claims that 'Europeans became brighter since antiquity, but then became stupider again since the nineteenth century.' At another he asks, 'are different races genetically predisposed to think about the world in slightly different ways?'

Eventually I ask what attracted Meisenberg – neither a geneticist nor a psychologist, but a biochemist working at a medical school – to this scientific niche. What got him so interested in race?

'I am not particularly obsessed with race,' he replies. 'But I got interested in the subject in the context of the question of why some countries are rich and others are poor.' He notes that there is 'a 50-fold difference in per capita GDP between the poorest countries in the world and the advanced Western countries' and he believes that learning ability, which he sees as being tightly linked to intelligence, is what makes the difference to a country's economic success. For him, this learning ability is programmed into a person's – or a population's – DNA. 'In consequence, the question of whether there are genetic ability differences between people in different countries is perhaps the most fundamental question in development economics.'

He tells me what a shame it would be for what he calls 'low-IQ countries' such as Pakistan to lose their brightest citizens if they emigrate to the West. 'This cripples the poor countries and makes it impossible for them to catch up,' he laments. If some nations don't have the cognitive ability to catch up with Europeans and East Asians, they will 'get stuck somewhere on the lower rungs of the developmental path'. At a stroke, he con-demns without evidence all the world outside Europe and parts of Asia to genetically inferior status.

Meisenberg certainly seems bothered by the social impli-cations of what he sees as immutable racial difference. He expresses a fear that the racial stock of smarter, wealthier nations is under threat and that the problem urgently needs to be addressed, particularly by immigration control. 'Populations that get too bright and too rich invariably slip into sub-replacement fertility and slowly breed themselves out of existence,' he writes, 'while those that are stuck at a lower economic and cognitive level also get stuck . . . with continuing high fertility of the less educated sections of the population.' These could so easily be the words of an early twentieth-century eugenicist. He, too, believes that the inferior might outbreed the superior.

*

The mystery for me as a journalist is not that scientific racists exist. There have always been those with prejudice of every persuasion in academia, and possibly there will always be. The bigger puzzle is how someone like Gerhard Meisenberg – a professor at a private university in the Caribbean – manages to keep the *Mankind Quarterly* afloat, and finds researchers to write the articles that fill its pages. This requires networks, it requires coordination, and it requires funding. The fact is, across the world, old-style scientific race research of the kind his journal publishes is deliberately discouraged by science funding agencies and governments, not to mention deeply frowned upon within academia. It is as controversial as it was when the journal was founded. To do it independently, even if only online and for a small readership, one needs resources.

Following the money is where the trail begins. Attached to one of his early emails, Meisenberg sends me a paper that attempts to describe how racial categories work. It was published in one of the 2017 editions of the *Mankind Quarterly*, authored by someone named John Fuerst, from an organisation called the Ulster Institute for Social Research. I have heard of neither him nor it. When I do a check, the institute doesn't appear in the UK's list of officially recognised higher education bodies, describing itself instead as 'a think tank for the support of research on social issues and the publication of works by selected authors in this field'.

It turns out that since 2015, the Ulster Institute for Social Research has been publishing the *Mankind Quarterly*, along with a handful of books on race. On the *Mankind Quarterly* website, its address is given as a postal box in London. Meisenberg himself sits on its advisory council. How the institute is funded, he won't elaborate; he reveals only that it operates on a shoestring budget. I find one link between the institute and an offshore company based in the Bahamas. 'There certainly is no regular external funding from any outside source that I know of,' he tells me. 'I guess it's more a situation where someone may donate a larger sum, perhaps as part of a legacy . . . That's how most of these small foundations work.'

According to a report in the *Independent* newspaper in 1994, the institute received $50,000 the previous year from the Pioneer Fund. It was a period in which the fund, then based in Manhattan, New York, was particularly active. An investigation by the *Los Angeles Times* around the same time estimated that it was dispensing roughly a million dollars a year to academics, most of whom the newspaper claimed were looking for genetic differences between races. As well as giving grants to scientists, psychologist William Tucker has noted, between 1982 and 2000 the fund handed almost $1.5 million to lobbying groups in favour of immigration reform in the United States.

If the Ulster Institute for Social Research really is operating on a shoestring today, part of the reason may be that the Pioneer Fund seems to have since declined to a standstill. 'It is my strong sense that it is not nearly as influential as it once was, largely due to the deaths of the fund's key players,' William Tucker tells me. The Southern Poverty Law Center, a team of civil rights lawyers based in Montgomery, Alabama, inform me that the Pioneer Fund has been mostly silent of late. As far as they can tell, in the last decade it has been gradually emptied of all its assets. In summer 2018 the Associated Press investigated tax records to find that it had given out nearly $7.8 million between 1998 and 2016.

Wherever their funds come from now, it is clear that there is a small cadre of researchers, some with very few academic credentials, who are still publishing and citing each other's research through organs such as the Ulster Institute for Social

Research and the *Mankind Quarterly*. It is a small, self-contained network that operates on the margins of respectability. The same names crop up again and again. Richard Lynn, the assistant editor of the *Mankind Quarterly*, is also president of the institute (one of its books is a tribute to him in his eightieth year). Lynn told the *Independent on Sunday* newspaper in 1990 that he had received grants from the Pioneer Fund. In 2001, he even published a history of the fund, titled *The Science of Human Diversity*. In his own investigations, Robert Wald Sussman wrote that Lynn 'does very little science and the "science" he does is extremely poor'.

Edward Dutton from Oulu University in Finland, the author of at least two of the institute's books, including one on racial difference in sporting ability, is also a regular contributor to the *Mankind Quarterly*. Another contributor, Tatu Vanhanen, a recently deceased Finnish political scientist, co-authored a 2002 book with Richard Lynn titled *IQ and the Wealth of Nations*. Vanhanen was inspired to enter this area of research after reading up on evolutionary biology, interpreting it as a way to explain inter-ethnic conflict. He believed that political ideology could be used to serve people's genetic interests, by keeping them loyal to their own ethnic group. In one high-profile magazine interview in 2004 Vanhanen claimed that the average IQ of Finns was 97, while in Africa it was between 60 and 70. The comment went on to cause a national scandal because Vanhanen's son had just become Prime Minister of Finland.

Whatever the scientific merits of their work, the researchers inside this group were and still are undeniably tight-knit. Most of them are largely unknown outside their circle but highly prolific within it. They have managed to build a thin veneer of scientific credibility that comes from getting published and cited, almost entirely by publishing and citing one another. And they keep finding new outlets for their work. The latest addition to this alliance is *Open Differential Psychology*, an open-access online journal that claims to have been set up in 2014 by a Danish research fellow at the Ulster Institute of Social Research called Emil Kirkegaard. It includes Gerhard Meisenberg and John Fuerst among its reviewers, and its published papers so far include studies of IQ in Sudan, and of crime among Dutch immigrant groups.

Even with mutual support, those who write for the *Mankind Quarterly* rarely make much impact outside the shadowy recesses of the Internet. But there have been a handful of higher-profile figures among them. Until his death in 2012, one was Canadian psychologist John Philippe Rushton, a former head of the Pioneer Fund and professor at the University of Western Ontario. Rushton became notorious in academic circles for claiming that brain and genital size were inversely related, making black people better endowed but less intelligent than whites. Despite this, he was important enough to have his work read and reviewed by genuine scientists.

One review in particular shone a spotlight on the kind of work that still routinely appears in the *Mankind Quarterly*. When Rushton's book *Race, Evolution and Behaviour* was published in 1994, the University of Washington psychologist David Barash was stirred to write, 'Bad science and virulent racial prejudice drip like pus from nearly every page of this despicable book.' Rushton, he added, seemed to have collected scraps of unreliable evidence in 'the pious hope that by combining numerous little turds of variously tainted data, one can obtain a valuable result'. The reality, Barash concluded, is that 'the outcome is merely a larger than average pile of shit.'

5

Race Realists

Making racism respectable again

In 1985, American historian Barry Mehler had a dream.

Now in his seventies and a professor of humanities at Ferris State University in Michigan, studying genocide, Mehler's investigations had in the 1980s taken him into the murky territory of academia's extreme right-wing fringes. His focus was on the founders of the *Mankind Quarterly* and Wickliffe Draper's notorious Pioneer Fund, both of which were now known to have been helping keep fringe elements in science alive for decades. As Mehler worked, he found his waking life began to soak into his subconscious, colouring his sleep. In his dream – in truth, more of a nightmare – his son, then around two years old, was trapped in a runaway car hurtling down a hill towards oblivion.

"The traffic is going in both directions, and I am in the middle of the road desperately waving my hands trying to stop the flow of traffic in order to save the life of my son,' he recalls. 'It's a dream. It's a metaphor for how I felt.'

A few years before, prompted by historical research he had already carried out into early twentieth-century American eugenicists and their links to Nazi Germany, Mehler had begun looking into what had happened to these same scientists and others with similar worldviews once the Second World War was over. Many people assumed that the eugenicists had all but disappeared with the Nazi regime, and that race science was essentially finished at the same time. What Mehler learned instead was that the prejudice that had existed before the war – the fear of some kind of threat to the 'white race' – was still alive in a few small intellectual circles.

'I was really focused on the ideological continuity between the old and the new, and the fact that these ideologies were malicious and dangerous,' he explains. What worried him most of all as he did his investigations was that these people seemed now to be stepping outside their own limited cabal to penetrate not just mainstream academia, but also politics. Their target was nothing short of the highest echelons of the United States government.

One of the key figures in the network to have survived the old days was Roger Pearson. Pearson's career trajectory was very different from that of Reginald Ruggles Gates, fellow founder of the *Mankind Quarterly*. During the Second World War, Pearson had been an officer in the British Indian Army. In the 1950s he worked as managing director of a group of tea gardens in what was then known as East Pakistan, now Bangladesh. And it was around that time that he began publishing newsletters, printed in India, exploring issues of race, science and immigration. Very quickly, says Mehler, he connected with like-minded thinkers all over the world. 'He really was beginning to organise, institutionally organise, the remnants of the pre-war academic scholars who were doing work on eugenics and race. The war had disrupted all of their careers, and after the war they were trying to re-establish themselves. Establishing these institutional networks was essential for their rehabilitation,' he explains.

Pearson's newsletters and the *Mankind Quarterly* relied on being able to reach out to marginal figures from all over the world, people whose views were generally unacceptable in the societies in which they lived. It was a job, of course, that was being done before the benefit of the Internet and social media, before it was easy for likeminded people with extreme views to easily find each other. 'You have these people who come seemingly out of nowhere. It was just so amazing to me that they would be so well networked,' says Mehler.

One of Pearson's publications was the *Northlander*, which described itself as a monthly review of 'pan-Nordic affairs' – by which it meant, more broadly and euphemistically, matters supposedly of interest to white northern Europeans. Its very first edition in 1958 complained about the illegitimate children born thanks to the stationing of 'Negro' troops in Germany after the war, and about immigrants arriving in Britain from the West Indies. 'Britain resounds to the sound and sight of primitive peoples and of jungle rhythms,' Pearson warned. 'Why cannot we see the rot that is taking place in Britain herself?' On the following page, he printed a tribute to Charles Darwin. He had made it his goal to awaken people to what he saw as the existential threat of immigration and racial intermixing, referring to anti-racists as 'cosmopolitans'.

Within a couple of decades, Pearson ended up in Washington, DC, establishing publications there, too, including the *Journal of Indo-European Studies* in 1973 and the *Journal of Social, Political and Economic Studies* in 1975. 'And that's what piqued my curiosity,' Mehler continues. 'Really looking at people who are racist at a time when liberalism was the predominant ideology.' In April 1982 a letter even arrived for Pearson from the White House, bearing the signature of President Ronald Reagan, praising him for promoting scholars who supported 'a free enterprise economy, a firm and consistent foreign policy and a strong national defense'. Pearson used this endorsement to help raise funds and support for his publications. It was clear that he and those in his circle must have had access to the very peak of the US government. The *Independent on Sunday's* 1990 investigation confirmed that Pearson received several grants from the Pioneer Fund around the same time.

Just as Mehler was carrying out his research, a soft-spoken civil servant in Washington, DC named Keith Hurt happened to be investigating the very same people in his spare time. When their paths crossed, Mehler and Hurt began combining their research. Hurt was then working for the Congressional Research Service, a branch of the Library of Congress that provides policy analysis to members of the House and Senate, which meant that he was keen to keep his identity private. Today, he tells me, he is free to talk on the record. 'I think I started out sort of naïvely,' he admits. 'I ran across some things that were disturbing, that I didn't expect. I didn't really understand that there were these structures and networks and associations of people that were attempting to keep alive a body of ideas that I had associated with at the very least the pre-civil rights movement in this country, and going back to the eugenics movement early in the last century. These ideas were still being developed and promulgated and promoted in discreet ways.'

Perhaps paradoxically, given the fierce nationalism of the figures involved, this also appeared to be a global network, spanning the United States and Europe at least, but also stretching to India and China. 'If you looked at the old *Mankind Quarterly*, it was a truly international journal, with contributors and editors from all over the world,' says Hurt. To this day, the *Mankind Quarterly* runs articles by many writers outside Europe, and its advisory board spans Russia, Japan, Saudi Arabia and Egypt.

'It was important to put together how the networks worked, where the funding came from, what the publications were, what the connections were. And what the connection is with right-wing political organisations,' adds Mehler. What he and Hurt were uncovering astounded them both. 'There was a network, an international network of these people who were not particularly well respected or regarded or even known outside of the network, but they had their own journals, their own publishing houses. They could review and comment upon each other's work,' explains Mehler. 'So it was almost like discovering this whole little world inside academia. And it was a rather nefarious world, of people whose origins went back to the Second World War.' What shocked them above all was the sheer professionalism of the operation, the slick ability of people with some of the most extreme views imaginable to connect to each other and communicate their views across thousands of miles. They were keeping scientific racism alive.

Mehler, who is Jewish, found it particularly disturbing. 'I have a lot of relatives who survived the Holocaust,' he tells me. 'When they flip the light switch and the light goes on, for them it's like "oh wow!". They are prepared for the world to collapse. They are prepared for things to cease to be normal very quickly because that was their experience.' I can hear the fear in his voice, an anxiety that political stability in even the strongest democracies sits ultimately on a precipice. 'I saw anti-Semitism. I was really alienated in American society. I was a person that felt that racism and anti-Semitism were predominant, and that the United States could easily become vicious, racist, and go back to its racist history when push came to shove, if people were threatened enough.' The past, he reminds me, is always capable of repeating itself.

It was around this time, as he and Hurt uncovered the network, that Mehler had his dream. 'I felt like I was desperately trying to prevent this from happening again . . . I thought that we were headed for more genocide.' The parallels between this far-right network of pseudoscientists and intellectuals and the rapid, devastating way in which eugenics research had been translated in Nazi Germany loomed large in Mehler's mind, terrifying him with the possibility that the brutal atrocities of the past could happen once more, that the ideological heart behind them was still beating.

*

Despite the urgency that both Barry Mehler and Keith Hurt felt, their investigations never made it into any high-profile publications. They appeared instead in a few small Jewish and left-wing newsletters, often with Hurt's name omitted or under an alias to protect his job at the Congressional Research Service. The lack of public interest reflected how many people assumed they no longer had anything to fear. Neo-Nazi political parties and white supremacists were thought to exist only on the irrelevant margins of real life.

'Race is such a difficult issue for Americans,' explains Hurt. 'People want to be optimistic. People want to believe that [racists] exist only in a sort of lunatic fringe, which is safely cabined off from the rest of society and that they have no consequences or implications for the future. That was the case then.' The world was thought to be moving in a liberal, more inclusive direction. Racists were thugs and skinheads, not men in power, not academics operating through covert networks.

Then in May 1988, Mehler and Hurt published an article in *The Nation*, a progressive weekly magazine, finally confirming that there might be reason to worry after all. It linked Ralph Scott, a professor of educational psychology at the University of Northern Iowa, to both the Pioneer Fund and the government. Their report claimed that Scott had used grants from the Pioneer Fund under a pseudonym in 1976 and 1977 to organise a national anti-busing campaign. Busing was a means of desegregating schools by transporting children from one area to another, and an important part of the civil rights movement. According to Mehler, some of Scott's money from the fund had also been used to sponsor a study in Mississippi looking at the physical and psychological traits of 'American Anglo-Saxon children'.

What turned the story into one of national significance was that in 1985 the Reagan administration appointed Scott the chair of the Iowa Advisory Commission on Civil Rights, a body whose express purpose was the enforcement of anti-discrimination legislation across the state. Just a few years earlier, Scott had brought, and later dropped, a lawsuit against three black civil rights activists who had described him as a racist. It was clear that even after taking up his influential post, Scott's views on racial difference hadn't changed. He continued to write pieces for the *Mankind Quarterly* and Pearson's *Journal of Social, Political and Economic Studies*. Indeed his most recent article for the *Mankind Quarterly* was published as recently as 2013. Scott, now an emeritus professor, refuses to give me any comments, or to confirm or deny Mehler's reports. But William Tucker has noted through his research that almost every one of his papers is a variation on the same theme: that integrated schools are holding back white students, and not improving achievement among black students, for the simple reason that the two groups are somehow genetically different.

So in 1985 here was Scott, a university professor known to be actively involved in blocking policies aimed at achieving desegregation, who had even been described as a racist, yet had somehow been made officially responsible for defending civil rights in his state. 'It was obviously alarming,' says Hurt. This also happened to come at a time when the Reagan administration was facing criticism for drastically cutting the Civil Rights Commission's budget, making his appointment look even more suspect. To outside observers, it was hard not to imagine that Scott had taken up the position as chair of the Commission to undermine it from within.

The month after Mehler and Hurt's article came out in *The Nation*, Ralph Scott resigned. At the time, although the story made some corners of the national press, it wasn't major headline news. 'It was largely dismissed, I would say, by people who in retrospect would probably admit [they] were mistaken to write it off,' says Hurt. Looking back on the case, in the context of today's politics, with the rise of far-right groups in Europe and the US, and of nationalism more globally, he believes that what they uncovered should have served as a warning. Scott was only an individual, but he operated within a larger network of intellectuals opposed to desegregation. 'What surprised me was how quickly and efficiently these groups worked,' adds Mehler. 'You would think it would be fringe people, and that they would remain on the fringe, and they would have difficulty raising funds and making contacts. That wasn't true at all. What surprised me was how quickly Roger Pearson went from Calcutta, India to Washington, DC to Ronald Reagan.'

Others within academia who have picked up the baton from Mehler and Hurt since then – including psychologist William Tucker at Rutgers University, who has carried out detailed investigations into the Pioneer Fund – have observed how well coordinated and resilient these networks have remained, even after key figures die off. When he set out to research the Pioneer Fund and its wealthy founder Wickliffe Draper in the late 1990s, Tucker tells me, 'I compiled a list of every academic or scientist I could think of who had been outspoken about racial differences and then searched the web or contacted their institution . . . Then I travelled to each of these places, fully expecting that some trips would be a waste of time and research money, because there would not be any Pioneer connection.' He was wrong. 'In fact, I never struck out. Every one of these persons had been contacted and usually supported either by Pioneer or by Draper.'

The Pioneer Fund may have since declined, but something important has entered to take its place. The global political landscape has changed, its focus moving away from the centre, making space once more for those at the extremes. The election of Republican President Donald Trump came at the same time as a growth in nationalist sentiment and far-right parties all over the world. For Hurt, the work he did three decades ago is prescient in today's political climate, not because Roger Pearson or Ralph Scott were ever particularly important figures in American politics during Reagan's time, but because they managed to gain close access to the government despite their views. Somehow, they both found a way to influence powerful people with their brand of intellectual racism.

'I've spent a lot of time in the last decade paying attention to the politics of immigration in this country, which are obviously related to all of that in intimate ways, and which dominate our politics in some ways today,' Hurt tells me. He believes what happened in the past can happen again. 'When Reagan came in, he didn't have established party networks of personnel that the establishment figures in the party had. So he cast a very wide net that included a very diverse range of people, including people like Ralph Scott. Scott wasn't, I think, representative of the central policy thrust of the administration, but there was a lot of carelessness at the beginning of the Reagan administration that allowed these people to step into positions of greater or lesser significance . . . He was symptomatic of a broader problem of entryism to the Republican Party by people like this.'

For those on the political extremes, it's a waiting game. As long as they can survive and maintain their networks, it's only a matter of time before society swings around and provides an entry point once more. The public assumed that scientific racism was dead, when in fact the racists were always active under the radar, suggests Hurt. 'I think there was a whole sequence of events between the late 1980s and the present in which these ideas, which have become pretty well established in the mainstream of American political culture, were step-by-step progressing, re-establishing themselves, eroding the norms of the post-civil rights environment.'

＊

What difference does it make to science that a publication like the *Mankind Quarterly* exists today? In truth, barely any. Its work is so rarely read or cited by mainstream scientists that its impact factor (the measure used to judge the influence of a journal) hovers between 0 and a little more than 1. By contrast, the impact factor of a highly respected journal like *Nature* is more than 40. But then, of course, the *Mankind Quarterly* was never designed to be read by scientists or shape the future of research. It was always a platform for those seeking intellectual ballast for their political views. What is of concern, then, is what it represents. As a publication, it's a barometer for intellectual racism. Should it or its contributors become popular, then we will know that something is wrong. And in the last ten years, its impact factor has been on average higher than it was in the preceding decade.

At the same time, its editors have built a presence in other, more credible scientific journals. Assistant editor Richard Lynn, for example, today sits on the editorial advisory board of *Personality and Individual Differences*, produced by Elsevier, one of the world's largest scientific publishers, which counts the highly respected journals *The Lancet* and *Cell* among its titles. Among Lynn's papers was one in 2004 on 'The Intelligence of American Jews', arguing that 'Jews have a higher average level of verbal intelligence than non-Jewish whites.' Gerhard Meisenberg's work, looking at the links between intelligence, genetics and geography, has appeared in *Intelligence*, a psychology journal also published by Elsevier. He has authored at least eight articles in recent years, including one in 2010 on the average IQ of sub-Saharan Africans, and another in 2013 on the relationship between 'national intelligence' and economic success. When I look at who's on the editorial board of *Intelligence* in 2017, I find both Meisenberg and Lynn listed.

While journals are free to publish whatever they think is worthy, subject to peer review, the choice of who to appoint to an editorial board is important because those members help to shape its policy and scope. According to Elsevier's own online guidance for editors, they 'should be appointed from key research institutes'. Neither Lynn nor Meisenberg can claim that honour. A spokesperson for Elsevier, after repeated prompting, tells me that editorial board members 'are not involved in making decisions about which articles will be published. Their role is focused on reflecting the academic debate that takes place within the communities' domain that the journal serves.' Yet Elsevier's own website states that editorial board members 'review submitted manuscripts' and 'attract new authors and submissions'. The other implication of their brief statement is that the work of Lynn and Meisenberg, studying population-level differences – which some might equate with racial differences – in intelligence must now be a part of mainstream academic debate.

In 2017, when I call the current editor-in-chief of *Intelligence*, Richard Haier, an emeritus professor in the medical school at the University of California, Irvine, to find out how he feels about having editors from the *Mankind Quarterly* on the editorial board of his own journal, he sounds nervous. 'I struggled with this, frankly, when I became editor, and I consulted several people about this,' he admits. 'I decided that it's better to deal with these things with sunlight and by inclusion.' Throughout our conversation, he is uneasy, taking long pauses to choose his words. Keeping them inside the fold, he tells me finally, reflects his 'commitment to academic freedom'.

Haier reassures me that he has never met Meisenberg or Lynn. But he tells me he did personally know and defend the late Arthur Jensen, a professor of educational psychology at the University of California, Berkeley. In 1969 Jensen mooted in the *Harvard Educational Review* that gaps in intelligence test results between black and white students might be down to genetics. His article remains one of the most controversial psychology papers ever published. The *New York Times* reported in

1977 that the Pioneer Fund had been subsidising Jensen's work. An investigation published by the *Los Angeles Times* almost two decades later confirmed that by then grants to Jensen from the fund must have totalled more than a million dollars.

Haier continues, 'The area of the relationship between intelligence and group differences is probably the most incendiary area in the whole of psychology. And some of the people who work in that area have said incendiary things . . . I have read some quotes, indirect quotes, that disturb me, but throwing people off an editorial board for expressing an opinion really puts us in dicey area. I prefer to let the papers and the data speak for themselves.' He adds that he personally does believe there is something scientifically interesting about studying 'group differences' in intelligence. 'Scientific intelligence research has laboured under this cloud for fifty years, and it is my stated goal as editor to help bring intelligence research back into the mainstream, where it used to be.'

Even so, Haier seems to have been bothered by my inquiries. When I check the Elsevier website around a year later, at the end of 2018, both Gerhard Meisenberg and Richard Lynn have been removed from the editorial board of *Intelligence*.

If group-level or population-level differences in intelligence do need to come out from under the cloud of controversy, then what are the reasons that researchers might want to wade into this deeply divisive area of research? According to Richard Haier, one theme is shared by some of those who submit their work to *Intelligence*. 'I can tell you, and I'm not revealing anything secret here as editor, we receive a number of papers that try to speak to the relationship between intelligence and economic development, and group differences in general.' Sadly, it's not always of the highest quality, he admits. 'When I read many of those papers, they are substandard, and they never even get to peer review . . . We have had papers submitted that come up with some kind of result, and then the discussion section extrapolates from that result to immigration policy. Those papers never get as far as peer review in our journal because it's clear that there is an agenda.' As I recall, these were exactly the kind of extrapolations made by Gerhard Meisenberg in his emails to me.

Haier's admission betrays just how much this field remains plagued by dark politics. I can't help but think of the nineteenth-century race scientists who jumped to biological explanations for the inequality they saw in the world, who believed other races had been doomed to failure by nature because their brains were too small or their temperaments too weak. Confronted by slavery and colonialism, they skimmed over history and culture, preferring instead to look to biology for justification for this kind of exploitation. When researchers like Meisenberg today link economic development to intelligence, they imply that the vast inequality between the world's richest and poorest countries is rooted not just in the imbalance of power or historical circumstance, but in the innate weaknesses of the populations themselves. Racial injustice and inequality, in their minds, isn't injustice or inequality at all. It's there because the racial hierarchy is real.

*

'I think what we're experiencing now is a much more threatening environment,' says Keith Hurt. 'We're in a much worse situation than we were a couple of decades ago.' He believes that the kind of research once sponsored by the Pioneer Fund and still published by the *Mankind Quarterly* has now found fresh avenues of support. Scientific racism has come out of the shadows, at least partly because wider society has made room for it. 'Frankly, I think at this point the ideological stream that it was sustaining is now self-sustaining. There are other institutions, and a much, much broader culture that will sustain it.'

Over the last thirty years this broader culture that Hurt describes, which rails against 'political correctness' and calls for a greater diversity of political opinion and freedom of speech in academia as a disguise for the propagation of extreme right-wing views, has become stronger. In 2018 an investigation by the *London Student* newspaper revealed that *Mankind Quarterly* editors Richard Lynn and Gerhard Meisenberg had been organising and speaking at a series of small, invitation-only conferences held at University College London since 2014. The Associated Press reported later that a University of Arizona psychology professor with an interest in human behaviour and evolutionary psychology, who has also been on the editorial advisory board of the *Mankind Quarterly*, had used a grant from the Pioneer Fund to attend one such conference. Somehow the organisers had managed to secure a space within the university to discuss controversial issues around eugenics and intelligence, attracting one attendee who was later appointed (although he hastily withdrew) to head the Office for Students, the regulatory authority for the English higher education sector.

Race scientists working two decades ago are building a stronger presence once more. In 1994, in *The Bell Curve*, one of the most notorious bestsellers of the twentieth century, American political scientist Charles Murray and psychologist Richard Herrnstein suggested that black Americans were less intelligent than whites and Asians. A review at the time in the *New York Review of Books* observed that they cited five articles from the *Mankind Quarterly*, and no fewer than seventeen researchers who had contributed to the journal. Murray and Herrnstein went so far as to describe Richard Lynn as 'a leading scholar of racial and ethnic differences'. Although *The Bell Curve* was widely panned after its publication, an article in *American Behavioral Scientist* describing it as 'fascist ideology', in 2017 *Scientific American* noted that Charles Murray was enjoying 'an unfortunate resurgence'. Facing down protesters, he was being invited to give lectures on college campuses across the United States.

Another contributor to the *Mankind Quarterly* has become a key figure in the white supremacist movement. Yale-educated Jared Taylor, who belongs to a number of right-wing groups and think tanks, founded the magazine *American Renaissance* in 1990. William Tucker describes it to me as today being the true intellectual arm of the modern neo-Nazi movement. One phrase Taylor uses to defend racial segregation, for example, borrowed from the zoologist Raymond Hall writing in the first ever issue of the *Mankind Quarterly*, is that 'two subspecies of the same species do not occur in the same geographic area'. His brand of white supremacy draws from race science to lend itself the illusion of intellectual backbone. Taylor is in some ways the Wickliffe Draper of the twenty-first century.

Like Draper, Taylor has sought to make racism respectable again. His American Renaissance Foundation conferences – which anthropologist Robert Wald Sussman has described as 'a gathering place for white supremacists, white nationalists, white separatists, neo-Nazis, Ku Klux Klan members, Holo-caust deniers, and eugenicists', and *The Journal of Blacks in Higher Education* has more succinctly dubbed 'a convocation of bigots' – have featured other intellectuals who have written for or edited the *Mankind Quarterly*. A visitor at the 1994 meeting reported that people didn't 'flinch from using terms such as "nigger" and "chink"'. Male attendees are expected to dress in smart business suits, to set themselves apart from the thuggish

image most people associate with racists. They don't call themselves racists but 'race realists', a euphemism that reflects how
they like to believe the scientific facts are on their side.

By the time of the 2016 US presidential election, Jared
Taylor's place in American politics was more secure than ever.
He even appeared in a television advert released by Hillary
Clinton's campaign team to show the kind of support that rival
Donald Trump's anti-immigration policies had among white
nationalists. Explaining the increasing prominence of people
such as Taylor during the election, one correspondent reflected
in *Newsweek* magazine in 2017: 'These men have degrees from
some of the nation's top universities . . . they are a well-read
group who cloak their ideas about the intrinsic superiority of
white men in selected passages of literature, history, philosophy
and science.'

One regular speaker at American Renaissance Foundation
conferences and another *Mankind Quarterly* contributor is
Michael Levin, a professor of philosophy at the City University
of New York (CUNY), which by its own mission as a large,
relatively affordable urban university is committed to being
accessible to under-represented minorities. He is the author of
Why Race Matters, a book which went out of print after publication in 1997, but was republished with a new cover in 2016,
together with a foreword by Jared Taylor. In 1986, Levin had
written a letter to the *New York Times* arguing that it was legitimate for store owners to discriminate against all black people
because they were more likely to be attacked by someone who
was black. According to the Southern Poverty Law Center,
Levin told the audience at a 1998 American Renaissance Foundation meeting, 'The two principal race differences that I see
are race differences in intelligence and in motivation . . . It's no
wonder there are very few black scientists . . . you have to have

an IQ of 130 to be a successful research scientist.' According to
an article in *The Journal of Blacks in Higher Education* in 1994,
Levin had by then received more than $120,000 in grants from
the Pioneer Fund.

'People like that are able to get funding,' Keith Hurt tells
me. 'There will always be, unfortunately, men of wealth – and
they are almost all *men* of wealth – who share these ideas and
are willing to support them.' But another reason that scientific
racists have more influence now is that the Internet and social
media have given them simpler ways to access and grow their
networks. 'Behind every racist joke is a scientific fact,' alt-right
blogger Milo Yiannopoulos told a Bloomberg reporter in 2016.
Among the cabal of 'race realists' operating online, as well as
their followers, it's easy to spot an attitude of doggedness. They
repeatedly insist that they are challenging the politically correct
wider world by standing up for good science, that those who
oppose them are irrational science-deniers.

Gerhard Meisenberg, for example, writes to me in our
correspondence that 'some academics seem to believe that by
simply claiming that race differences don't exist, we can prevent
people from believing in them. Doesn't work like that . . . For
example, if we tell people that black children are as smart as
white children and it isn't true, there will be teachers and others
who know first-hand that it isn't true. Also, if we tell people that
it's because of some flaws in the school system and the flaws
are repaired and it doesn't help one bit, not only do people
get frustrated about all the wasted effort, but they also start dis-
trusting the "scientists" who are telling them lies. Real people
aren't postmodernists. They distinguish between truth and lies.'
Meisenberg's tactic is simple: using people's gut prejudices and
casual observations to undermine trust in mainstream science.
If you *feel* it to be true, it must be.

This rhetoric around who has the genuine claim on the truth resonates today more than ever when the public on both sides of the political divide worry about fake news and media conspiracies. Yet it's a line that has been adopted by intellectual racists for decades. In 1998 an American corporate lawyer called Harry Weyher, president of the Pioneer Fund from 1958 until he died in 2002, was given space to write an eighteen-page editorial for the journal *Intelligence*. He used the opportunity to defend research supported by the fund. 'This is critical research by world-class scholars . . . Yet, if one were to believe some important segments of the media, this research was funded by an evil foundation and done by evil scientists, and is unfit for public dissemination.' Twenty years before Donald Trump was elected President, Weyher too laid into what he saw as an egalitarian orthodoxy and political correctness, going so far as to accuse the print and broadcast media of 'false reporting'.

*

'Why do we still have race science given everything that happened in the twentieth century?' I'm asked by American anthropologist Jonathan Marks, an academic many turn to for clarity when it comes to racism in science.

His answer is unequivocal: 'Because it is an important political issue. And there are powerful forces on the right that fund research into studying human differences with the goal of establishing those differences as a basis of inequalities.' Ultimately, politics is always a feature of the science, just as it was in the very beginning. Once there was the backdrop of slavery and colonialism, then it was immigration and segregation, and now it is the right-wing agenda of this age. Nativism remains an issue, but there is also a backlash against greater efforts to promote racial equality in multicultural societies. And just like before, the message of those with racist intentions is tailored gently, carefully sculpted to appeal to populist fears while at the same time sounding logical and reasonable. Communicating with me, for example, Gerhard Meisenberg uses the word 'culture' alongside the word 'race' as though they're fully interchangeable, understanding how much most people these days value and respect cultural boundaries, even if they don't recognise biological race.

'Without much selectivity in migration, all countries of the world become homogenised, not only in ability level but also in culture and everything else. Countries become more similar to each other,' writes Meisenberg. How tragic it would be to have the whole world look exactly the same. On a logical level he fails to explain how, if races are fixed and immutable in the way he thinks they are, we could all end up the same just by migrating. But this isn't the point. On the surface, heard quickly, his concerns sound almost sensible, in the same way that eugenics sounded so rational and attractive to progressive social reformers in the early twentieth century. Who back then could argue against the pursuit of a healthier, stronger population? Who today could argue against countries and groups maintaining their distinct cultures?

Race has always been an intrinsically political area of research, the idea itself born out of a certain world order. So it's small surprise that those doing this kind of research end up in the same place again and again, as they have for centuries. When they look for human variation, however objective they claim to be, they can't help but ask what the differences they think they see mean for society. This is why race science so often comes twinned with speculation as to the causes of economic and social inequality.

A common theme among today's race realists is their belief that, because racial differences exist, diversity and equal opportunity programmes – designed to make society fairer – are doomed to fail. 'As far as I understand, race-based policies of this kind were adopted by many American institutions since the 1960s or 70s, and were originally justified as something that is needed to make up for disadvantages that Blacks and some other minorities had suffered in the era before civil rights legislation,' Meisenberg tells me. 'Today, fifty years later, that old reasoning is no longer credible.' Rather than investing in these policies, then, he appears to argue that we should accept inequality as a biological fact. If an equal world isn't being forged fast enough, the race realists don't see it as a longer or tougher path than we imagined. Instead, it is a permanent natural roadblock created by the fact that, deep down, we're not the same.

'We have two nested fallacies here,' Jonathan Marks continues. The first is that the human species comes packaged up in a small number of discrete races, each with their own different traits. 'Second is the idea that there are innate explanations for political and economic inequality. And basically what you're doing there is saying that inequality exists, but it doesn't represent historical injustice. What these guys are trying to do is manipulate science to construct imaginary boundaries to social progress.'

For Marks, the prescription for this pathology is radical. In science, to ban any kind of research – if it can get funded and ethically approved – is a risk to academic freedom. He suggests instead that people who 'cannot handle the results shouldn't be studying it . . . We don't want racists working on human variation because that doesn't work. So it's not a question of "should this be studied?", it's a question of who should be studying it and how should they be credentialled, or how should they be vetted.' His argument is that if this field is always going to be affected by the politics of the scientists, then surely it makes sense to have people doing the research who aren't bent on division and destruction, whose aims don't lie at odds with what society as a whole has decided is morally acceptable.

To others, this sounds heavy-handed. Psychologist William Tucker, for example, whose painstaking work helped expose the Pioneer Fund and the scientists it has supported, supports the freedom of anyone to do any research – even the kind he abhors. 'People enjoy the right to take Pioneer's money; I would not like to see them deprived of that right. At the same time, I think that their decision to do so is awful and will do whatever is in my power to persuade them of the folly of such a course,' he tells me.

Fundamentally, though, the problem is not the science itself. If it were limited to academia, the kind of material that's published in the Mankind Quarterly and other likeminded publications would have next to no impact at all because mainstream scientists almost entirely ignore it. It just doesn't have enough scientific value. As for the ideology, most researchers today accept that what little we know about human variation can't be used to dictate how we treat people in the real world. It certainly can't be used to set policy. The problem is in how these ideas are used and abused in wider society, how much traction they can gain with the public and those in power. Nazi scientists carrying out their regime's programme of 'racial hygiene' had only a rickety scientific framework upon which to justify the destruction of millions. The same was true of those who used science to defend slavery, colonialism and segregation. And it applies today, too, to those on the political extremes. Nothing is more seductive than a nice string of data, a single bell curve, or a seemingly peer-reviewed scientific study. After all, it can't be racist if it is a 'fact'.

For those with a political ideology to sell, the science (such as it is) becomes a prop. The data itself doesn't matter so much as how it can be spun. Marks warns me, then, that those to really watch out for are the ones who claim to be uniquely free of bias, who tell you they have some kind of special, impartial claim on the truth. 'Whenever anybody tells you I am objective, I am apolitical, that is the time to watch your wallet. Because you're about to have your pocket picked.'

And he should know, because he almost had his pocket picked.

6

Human Biodiversity

*How race was rebranded for the
twenty-first century*

IT WAS 1998 or thereabouts that an invitation arrived in Jonathan Marks' email inbox. He doesn't recall the precise date
because, until I asked him about it, he hadn't thought about it
for years. What he does remember is that the sender was a then
little-known American science journalist and former writer for
the conservative *National Review* magazine, Steve Sailer. The
invitation was for Marks to join a mailing list of people interested in the subject of human variation.

'I knew absolutely nothing about him,' Marks recalls. 'He
just seemed to be someone who was organising something.'

At the time, Marks was teaching at the University of
California, Berkeley, a few years after writing a popular text-
book on race, genes and culture titled *Human Biodiversity*.
Wedging these two words together, he had neatly coined a
phrase to describe biological and social variation across the
human race. Part of the reason he chose it, he admits, is that
'biodiversity' had become something of a buzzword. He never

guessed it would cause any problems. And why would it? Diversity was being celebrated; both the glorious biodiversity of the natural world and the rainbow of cultural and physical diversity in human societies. It was the proud label of liberal anti-racists, of the good guys like himself.

'America's answer to the intolerant man is diversity,' Robert F. Kennedy had affirmed in 1964 at the dedication of an interfaith chapel in Georgia. 'United in diversity' was the motto of the European Union. It goes without saying that although different cultures have different things to offer, the Benetton breadth of human variation does nothing to undermine the general consensus that we're biologically pretty much all the same beneath the skin. This is a truth that's been universally acknowledged since the end of the Second World War. At least that's what Marks thought.

The invitation itself appeared perfectly innocent. The 1990s marked the early years of electronic mailing lists, and Steve Sailer apparently wanted to use one as a way of pulling together scientists, intellectuals and fellow journalists to start a private conversation about human difference. 'He said, hey, I'm interested in human variation, and I like your work. Let's get an email list together,' says Marks. 'It seemed pretty straightforward and harmless.' So he signed up. What intrigued him especially was that Sailer happened to be brandishing Marks' own neologism, calling his list the 'Human Biodiversity Discussion Group'.

Others joined in their dozens. By the summer of 1999, Sailer's roster of members was astounding. Along with prominent anthropologists like Marks, there was psychologist Steven Pinker, political scientist Francis Fukuyama and economist Paul Krugman. In hindsight, the large number of economists in the group might have been a warning. There in the mix, too, was the controversial author of *The Bell Curve*, political scientist Charles Murray. That should have been more of a red flag.

When Marks had talked about biological diversity, he meant the superficial variation we see right across the species, from individual to individual. 'There isn't really room for three, or four, or five biologically distinct kinds of people,' he tells me. He certainly didn't expect to see people reinforcing old-fashioned stereotypes of the kind that had long been debunked. That school of racism was long dead, he assumed. Yet here on this email list, something strange was happening. Observing the conversations that Sailer steered through the group, Marks noticed the term 'human biodiversity' being used differently from the way he had originally intended. Members were using it to refer to deep differences between human population groups.

Among the people added to the group that summer was Ron Unz, a Harvard graduate and founder of a financial services software company, who had recently run as the Republican candidate for governor of California. His introduction to other members was pasted alongside a 1994 article he had written in *The Wall Street Journal* titled 'How to Grab the Immigration Issue', observing the state of California's changing racial demographics. 'Conservatives and Anglos have become enormously angry and frustrated over the growth of crime, welfare, affirmative action and the general decay of their society,' he stated. Another addition to the group was Deepak Lal, a professor of international development at the University of California, Los Angeles, whose work explored the reasons the West happened to be economically more successful.

It dawned on Marks that Sailer's seemingly innocent email list was not so much about discussing science in an objective way, but more about tying together science and economics

with existing racial stereotypes. One debate that sticks in Marks'
mind today took place between himself and a journalist who
claimed that black people were genetically endowed to be
better at sports. Marks insisted that this was a scientifically
shaky argument, not to mention one with dangerous political
implications. The two experts clearly disagreed. But rather than
help reach a consensus, 'Steve Sailer clearly took his side,' he
tells me.

'At which point I realised, ah! This isn't an impartial
scholarly discussion.'

Another time, Sailer defended a writer on the list who suggested that different ethnic groups had their own particular
strengths, adding that Turks were born physically more powerful. This was, he suggested, one reason why affirmative action
policies to hasten racial equality weren't a good idea. Different
ethnicities should instead be encouraged to do what they do
best. When Sailer talked about human biodiversity, he didn't
appear to be using the phrase in a politically neutral way, but
as a euphemism. He had spun the same language used by
liberal anti-racists to celebrate human cultural diversity in order
to build a new and ostensibly more acceptable language around
biological race.

In email correspondence with me Sailer denies duping
anyone, although he does admit that the group leant towards
the heretical. But whatever his intentions, this wouldn't have
been the first time someone had tried to subvert the idea of
'diversity'. In their conclusion to *The Bell Curve*, a book which
just a few years earlier had claimed that black Americans
were innately less intelligent than whites, Charles Murray
and Richard Herrnstein similarly undermined the political
push towards racial equality by arguing that biological differences between groups made it practically impossible. 'We are

enthusiastic about diversity – the rich, unending diversity that free human beings generate as a matter of course, not the imposed diversity of group quotas,' they wrote. Every person in a diverse society had a valued place, they implied – just not the same place.

Realising his error, Marks left the group. Other members also denounced it as racist. 'If somebody said the same to me today I would probably be a little bit more suspicious,' Marks admits. 'I would look at who else is on the invite list. But even that can be misleading because he may be inviting people who are just as confused about this as I am.' At the time, it seemed harmless. Just a bunch of people with some marginal political ideas trying to convince others of the same.

Instead, Marks was about to find out just how prophetic the existence of the list was. 'That was my introduction to what became the alt-right.'

*

For those sucked into Sailer's electronic arena for the intellectual discussion of race, his email list was just a taste of the virulent racism that would later be seen far more often in shadowy corners of the Internet, then more openly on social media and right-wing websites, and finally in mainstream political discourse. Many more soon took hold of the phrase 'human biodiversity', giving it a life of its own online. Today it's nothing short of a mantra among self-styled race realists.

'Blogs have made the dissemination of wacko ideas much more efficient,' says Jonathan Marks. 'Actually in academia, we don't really know what to do with blogs because they have a really short shelf life. You forget about them a day after you read them, it's hard to cite them. But today, they are out there. They're these markers of people with wacko ideas.'

To be fair, few could have guessed that the email list was a precursor to something bigger. But as the group slowly became defunct, Steve Sailer's political convictions became increasingly obvious. He and other members of the list went on to become prominent conservative bloggers, writing frequently on race, genetics and intelligence. As a columnist for VDARE.com, an American website that describes itself as a news outlet for patriotic immigration reform (and, like the *Mankind Quarterly*, is automatically blocked by my Internet service provider), Sailer once argued passionately for the biological reality of race, stating that it was a fundamental aspect of the human condition. In 2009 the same right-wing website published Sailer's first book, about Barack Obama, titled *America's Half-Blood Prince*. In 2013 Ron Unz founded his own blogging platform, the *Unz Review*, as an alternative to the mainstream media, recruiting Sailer as one of his most prolific columnists.

Sailer truly rose to prominence, though, in the United States presidential election of 2016. Six years earlier, he had proposed focusing heavily on immigration to draw in white working-class voters as a single bloc. Having subverted the language of diversity through his email list, he did the same with identity politics. If ethnic minorities, such as black and Hispanic Americans, could assert their rights and defend their interests, he reasoned, then why not white voters who felt they were losing out to cheap immigrant labour and globalisation? It turned out to be an unexpectedly successful campaign strategy for Donald Trump.

'I think the big problem this country has is being politically correct,' Trump said in the first Republican Party presidential debate, in Ohio in 2015. Dog-whistle politics was reframed as a pushback to liberal elitism. A similar approach was adopted by some campaigners in the UK in favour of leaving Europe during the Brexit debate around the same time, emphasising that Turkey might join the European Union and flood Britain with millions of Turkish migrants.

An article in *New York* magazine in 2017 described Sailer's string of prophetic political insights as a new wave of populist thinking on the right, dubbing it 'Sailerism'. In the United States at least, he was credited with inventing a form of identity politics for disgruntled poorer whites, a group that had been neglected by politicians. But at the same time, political observers couldn't help but notice that his ideology often looked suspiciously like white nationalism.

For researcher Keith Hurt in Washington, DC, who has kept a close eye on the immigration debate ever since his investigations into right-wing intellectual racism in the 1980s, none of what happened during that time should have come as a surprise. 'The election of Trump made it impossible for many people to any longer overlook this stuff,' he tells me.

Hurt explains that the racist ideologies that existed at the start of the twentieth century, manifesting themselves in the eugenics movement and then in German nationalism, had survived by the end of it. The only difference was that those who held these views were later forced to keep them private. 'The post-World War Two ideological consensus rested on an implicit social contract that said, on the one hand, overt racist language will not be deployed in the public square in the way it was before. On the other hand, if this kind of talk is pushed out of the public square, society will refrain from making accusations of racism against all but the most extreme fringe.' And so the intellectual racists, the ideologues, communicated with each other through their own tight networks, disseminated their ideas through their own publications, some of which were so marginal and private that they were almost invisible to the outside world. It was easy for the public to assume they didn't exist, that the only genuine racists left were the ones they could see and hear, the skinheads and thugs.

When the time was right, however, political and intellectual racism slowly resurfaced into the mainstream. Sensing the rising tide, people like Steve Sailer waded in. 'They couldn't exist as they do now if there hadn't been the intellectual ideological continuity,' adds Hurt. They weren't starting from scratch, they were just repackaging the old ideas that they had been nursing the entire time, the same racial assumptions that had been around for decades.

But it all came as more of a surprise to academics like Jonathan Marks. 'I was working with the assumption that these guys were a lunatic fringe. If you had told me twenty years later that they would be part of a political mainstream wave, I would have said you are absolutely crazy. These guys are anti-science. These guys are positioning themselves against the empirical study of human variation and they are clearly ideologues for whom empirical evidence isn't important,' he laughs.

'But I think they were a lot cleverer than us professors.'

*

Offline, in mainstream academia and among respected scientists, a different debate was taking place. But it would similarly come to reshape the way people thought about race.

It was 1991, shortly after the launch of the multi-billion-dollar Human Genome Project, which was steadily working to build a map of the genetic data shared by our species for the first time by sequencing the entire length of human DNA. At the Stanford University School of Medicine in California, an influential geneticist by the name of Luigi Luca Cavalli-Sforza, then around seventy years old, spotted an opportunity. His own long and illustrious career had been dedicated to studying human variation across the world, work that had taken him from Italy to the United States and brought him to the forefront of a field known as population genetics.

Cavalli-Sforza and some of his colleagues – a small team of anthropologists and geneticists, based mostly in the United States – wondered whether the kind of data being collected by the Human Genome Project could also be used to pick through the fundamental differences between human groups, the genetic variations that make us who we are. Little did they guess that it would turn out to be one of the most controversial scientific initiatives of its time, leading to decades of debate about the biological reality of race.

Unlike Steve Sailer, Cavalli-Sforza and his team didn't come to their new project with race in mind. Indeed, quite the opposite. They were avowed anti-racists, fully signed up to the UNESCO statements on race in the 1950s, and firm believers that science would do nothing but prove racial stereotypes incorrect. They thought that their work might even help free the world from the scourge of racism. 'This was, at least on the American side, in general political terms, a quite left-wing group of people,' recalls Henry Greely, a professor at Stanford Law School who became involved in the initiative later on.

Cavalli-Sforza was used to receiving hate mail from people who disagreed with his outspoken belief that genetics didn't support old-fashioned notions of race. In 1973 he had publicly debated with William Shockley, a Stanford University physicist and joint Nobel Prize winner, who in later life became a notorious racist. Shockley believed that black Americans had intellectual shortcomings that were hereditary, and that black women should therefore be voluntarily sterilised. He was among the most prominent race theorists to receive support from the Pioneer Fund. When they met at Stanford, Cavalli-Sforza coolly demolished his claims, fact by fact.

Population genetics itself was born out of post-Second World War efforts to move away from traditional race science and eugenics. In the 1950s and 60s, when geneticists stopped talking about race, they turned their attention instead to 'populations', the 'human variation' between these populations, and the 'frequencies' of certain genes within these groups. It was a more rigorous, molecular, mathematical approach to studying human difference. And what population geneticists like Cavalli-Sforza quickly noticed was that there are no hard genetic boundaries around human groups, but rather continuous statistical variation, with a good deal of overlap. What differences there are exist along gradients, not borders. But that said, variation isn't random either. It can, depending on how you look at it, fall into clusters in which certain genes are statistically more common in some groups than in others.

Cavalli-Sforza and his peers became fascinated by these clusters. It was thought they would be more obvious in places where people had been geographically isolated for hundreds of years, living on islands or at the tops of mountains, mating almost exclusively within their own group through many generations. 'Primitive groups' were believed to be particularly distinct genetically because of the length of time they had spent away from others, shielding their genomes from the effects of intermixing. Since these people were so remote, their lineage conspicuous thanks to their long seclusion, maybe their genes could offer special insights into how humans evolved and adapted. Cavalli-Sforza was among those who believed that, by studying the genomes of these primitive groups, it might be possible to track historic patterns of migration. If they could see how gene frequencies varied in such groups and their neighbours, they could perhaps track where their ancestors lived in the distant past. In 1961 he became one of the first scientists to apply modern statistical methods to see how frequencies of major blood types varied among large human groups, creating a family tree to then show how these groups were related.

The search for what they thought would be genuine difference at the corners of the world took researchers to the remotest places imaginable. In 1964 the World Health Organisation picked out 'Eskimos' in the Arctic, Guayaki hunter-gatherers in the forests of eastern Paraguay, 'pygmies' in the Central African Republic, Aboriginal Australians, and the tribes of the Andaman Islands in India as potentially interesting 'primitive groups'. Scientists who couldn't make the trips would settle on immigrant groups in their own backyards, particularly Jewish and Roma communities, also known to be ancient and tight-knit. There was no point in studying very large human populations, say Africans or Europeans, because there was too much variation within them as a result of migration and mixing. Small, old indigenous communities were thought to be special, more distinct.

On the back of this research, Cavalli-Sforza formed a plan. He mooted the idea of using the same revolutionary gene sequencing technology as the Human Genome Project not just to map one genome, but by travelling the world to draw genetic data from lots of individuals of different ethnicities. This 'Human Genome Diversity Project', he and his collaborators announced in the journal *Genomics* in 1991, would 'supplement and strengthen findings from archaeology, linguistics, and history'. Through their work, genetics would sit alongside the study of culture, language and history to help paint the human story.

The Human Genome Diversity Project would initially zoom in on four to five hundred different small populations, especially geographically remote and apparently dwindling ones. Referring to these indigenous groups as 'isolates' – people including the Basques in Europe, the Kurds of eastern Turkey, and Native Americans – they expected their genetic data to provide clear signals about prehistoric migration and social structure. Calling for urgent funding from the world's governments, their argument was that this was a race against time to document the breadth of human variety before we all fell into the great big melting pot, when migration and assimilation would leave us all so genetically similar that such an effort might not be quite so worthwhile.

But what the scientists were proposing didn't sit easily with everyone. Some outside observers couldn't help but be a little uncomfortable. After all, it was hard not to wonder whether in the nineteenth century, this might have just been called race science. The Human Genome Diversity Project was technically more precise, of course, more scientific. It wasn't sampling skin and hair colour, or slotting people into racial hierarchies. It was using genetics. But in some ways it was hardly distinguishable from the study of human difference a hundred years earlier. The word 'race' had been prudently replaced by 'population', and 'racial difference' by 'human variation', but didn't it look suspiciously like the same old creature?

Then again, could it be called race science if the scientists involved were obviously anti-racists? As Henry Greely told me, they were left-wing liberals committed to stopping racism, whose public lives had been dedicated to fighting scientific racists and eugenicists. How could there be anything to worry about?

*

'They won't use the term "race",' I'm told by Joanna Radin in the department of history at Yale University, who has studied the progress of the Human Genome Diversity Project since 1991 to the present day. She notes that there was always a deliberate effort to avoid the word 'race', mostly because of its obvious political baggage, but also because the scientists didn't see genetic 'isolates' as being 'races' in the traditional sense. They were different enough to merit study, but not in the way racial difference had been described in the past. They were small groups, not large continental-scale ones. 'They don't necessarily map on to existing racial hierarchies,' she says.

Those behind the project insisted that their research was countering racial myths. Their stated intention was to replace ignorance and prejudice with hard scientific facts, and make it clear that we are one single human species, united in our common origins. Their plan was to look for difference not as racists had in the nineteenth century, to prove inferiority or superiority, but to use the tiny difference deep in our genomes to help build a picture of human migration. They would be like archaeologists, digging through our genes, looking for clues about our history. Their aim was simply to understand our past.

Given the unimpeachable political credentials of people like Luigi Luca Cavalli-Sforza, the scientists behind the Human Genome Diversity Project might have expected it to go forward without a hitch. All they needed was funding, and permission from their 'isolates', the indigenous communities whose blood and DNA they wanted to sample. But things weren't so easy. This wasn't the sixties any more, when foreign researchers could pick the communities they wanted to study and just rock up in the expectation that they would comply. Those communities had become more wary.

'Luca Cavalli-Sforza was an old school anthropologist,' geneticist Mark Jobling at the University of Leicester explains. 'I mean, I went to talks by him in the nineties where he would show old slides of him collecting DNA, blood samples in Africa from pygmy groups and offering glass beads and cigarettes in return, things like that.' This wasn't the way things were done now. This was the 1990s, the dawn of the Internet age, identity politics, and the fight for indigenous rights. Scientists who took an interest in indigenous communities now found that the same communities were taking an interest right back. They weren't as trusting as they had been in the past. And they were organised.

And they had good reason to be organised. Remote tribes and ethnic groups had been exploited throughout history, their land and cultural artefacts pillaged by Western colonisers, their bodies targets for unethical experimentation. Between 1946 and 1948, for example, the United States government ran secret experiments on thousands of people in Guatemala, deliberately exposing them to sexually transmitted diseases. Before and during the Second World War, British scientists had deliberately sent Indian soldiers fighting for Britain into gas chambers to study their response to mustard gas exposure. There had been a long and bloody tradition of scientists abusing other populations for their own ends, particularly populations deemed at the time to be racially inferior. But people were now prepared. Rights activists, alert to the risk of exploitation, were ready to defend the communities targeted by the Human Genome Diversity Project.

These activists warned of the possibility that DNA analysis might damage how the communities concerned chose to understand their past, reveal something profitable, or even be used as a weapon against them by racists. They weren't prepared to hand over their biological data, their blood and tissue, know-ing that it might end up being misused. Even so, their resistance to join in the project baffled some of the scientists involved. As Radin explains, 'What had changed – and this is what caught the scientists by surprise – was that the indigenous groups they had imagined disappearing, that they didn't have to reckon with once they left, these purported isolates, were organising in indigenous movements. They had access to the web, they were in touch with activists.'

As the future of the project came increasingly under threat, Cavalli-Sforza and his colleagues called in Henry Greely from Stanford Law School to help navigate the ethical dilemmas and deal with critics. Greely, fascinated by the project and taking a personal liking to the people inside it, agreed. At the begin-ning, he had very little knowledge of the science itself. 'I knew how to spell DNA but that was about it.' But from the outset, he could see that there were likely to be problems. The scien-tists were of course aware that science did not have a great historical record when it came to race. 'They knew this was an issue . . . they knew that it had a bad past.' But they didn't see themselves as part of that past. As far as they saw it, 'they were the *break* from the past. They were the good guys. They were the ones who understood the rights-enhancing, equality-enhancing potential of genomics, and they were going to bring it to the world,' he says. Greely adds that this was the first place he came across the fact 'that all humanity is more similar to each other than a band of chimpanzees that lives in a particular region of Africa'.

But at the same time, they were oddly naïve about how the project might be seen from the outside. One scientist 'talked about the need to sample "isolates of historical interest", a term that indigenous populations did not care for,' Greely admits. 'It struck me that that was not likely to be well received because it's a very clinical, bloodless way of referring to people who are alive, and cultures that are living now. Historical interest is sort of something you find in a museum. It was tone deaf.'

'Naïveté is always easily diagnosed through the retrospecto-scope,' he tells me.

Greely's job, to navigate the ethical quandaries posed by the Human Genome Diversity Project, turned out to be a poisoned chalice. Partly because of the political controversy surrounding the project, it didn't attract the funding they wanted. Yet the scientists struggled to understand why. 'Some were so comfort-able with their own knowledge of their own moral bona fides that it was hard for them to imagine being attacked from the left. They would have imagined that any remaining *racists* would be attacking them, not that *they* would be attacked as being racists.' If anything, they were on the same side as their attackers. 'I was the most conservative person on the North American commit-tee, and I'm a Carter-Clinton-Obama Democrat!'

The activists representing indigenous groups turned up the dial on their protests. At a meeting of the World Council of Indigenous Peoples in Guatemala in 1993, Greely found him-self facing down the accusation that he was a CIA agent, intent on committing genocide, which was clearly untrue. 'We took some lumps. Some of them were deserved and some of them weren't.'

In an address at a special meeting of UNESCO in 1994, Cavalli-Sforza turned his focus to the charge of racism, insisting that his project would help combat prejudice, not perpetuate it. But in 1995 another political storm blew up when scien-tists funded by the United States National Institutes of Health tried to patent a virus-infected cell line from people belonging to the Hagahai tribe in the highlands of Papua New Guinea for the purpose of developing a new treatment for leukaemia. Activists accused them of stealing people's biological samples and attempting to profit from them.

It was in this charged political environment that Greely drew up a model ethical protocol, setting in stone that samples wouldn't be taken unless entire groups, not just individuals, had given their consent. But despite such concessions, there was a deeper problem. At the same time that the project claimed to be anti-racist, it was hard to escape the paradox that this was also all about finding out how people differed. If the genetic variation between us was already known to be trivial, then why embark on a multi-million-dollar international project to study it at all? In what way did this reinforce that we were all the same underneath?

For geneticist and critic Mark Jobling, the way the project was structured, deliberately going after isolated populations rather than scanning people all over the world wherever they happened to be, was what ultimately undermined it. 'How you define the population in the first place, these are culturally loaded things in themselves. So there was a lot of cultural discrimination in the original aims.' The isolated communities that scientists such as Cavalli-Sforza believed were unique were actually never all that isolated or unique, but they were treated as though they were.

When I raise this with him, Greely admits that there was an 'uneasy recognition that if your project is about looking for differences then it's sort of counterintuitive to say it's showing similarities'. Anti-racist as the scientists behind the project were, they had somehow fallen into the trap of treating groups of people as special and distinct, in the same way that racists do. They were still forcing humans into groups, even if they weren't calling those groups races. They were using similar intellectual frameworks to pre-war race scientists, but with fresh terminology.

That's not to suggest that Cavalli-Sforza was ever duplicitous or secretly racist. Perhaps he saw human differences as meaningful, but didn't want to focus on that uncomfortable aspect of his work in a world in which drawing attention to it could have political consequences. 'I don't think he believed himself to be engaged in a racist enterprise,' says Joanna Radin at Yale University. 'But I also don't think he really had any more sophisticated a sense of how this was going to fight racism than just being able to show we're all connected, we're all cousins or something.' Jobling adds that the anti-racist aims of the project weren't mirrored in the structure of the project itself. Anti-racism seemed to be more of a political ideal tagged on as an afterthought. 'They did have a slightly happy-clappy narrative to it, you know, joining up the whole human family kind of thing.'

In reality, it would have been perfectly possible to study human variation without grouping people. As Jobling explains, the divisions between us are so blurry that humans can theoretically be grouped any way you like. 'You could do a thought experiment where you just said we will take Kenyans, Swedes and Japanese, and will just proportion everybody into those three things.' If this were done, because we are all genetically connected to the average Kenyan, Swede or Japanese person, either directly or by historic migration, then everyone on earth could theoretically be fully grouped based on just these three nationalities. 'You could say that you were so many per cent Kenyan, so many per cent Swedish, and so many per cent Japanese.' This may seem meaningless, but actually it is no more so than dividing the world into black, brown, yellow, red and white. 'The definitions of those populations are cultural, and the choice of population is driven by expediency.'

Other geneticists have also warned against dividing up the world this way. It imposes a certain order on our species, ignoring the actual fuzziness. If the Human Genome Diversity Project had proposed sampling people more systematically, in a grid pattern across the globe perhaps, the true overlapping nature of human variation would be easier to see. Scientists would have been able to map gradual, continuous variation across regions, rather than tight knots centred on very small communities. It's hard not to imagine that this approach – which in fact was mooted at the time, but then discarded – might also have been a more effective way of fighting racism. But in the end, it wasn't the one the researchers chose.

Most governments, including that of the United States, proved unwilling to invest in the Human Genome Diversity Project. It never quite got off the ground in the way it was envisioned. To this day, it remains something of a cautionary tale. In hindsight, part of the problem was that the scientists, however well intentioned they were, failed to connect what they were doing with people's real-life experience of race, with the history and politics of this deadly idea. They thought they were above it all, when in fact they were always central to it.

*

Luigi Luca Cavalli-Sforza died in 2018 aged ninety-six. When I emailed him shortly before his death, he was retired and living in Milan. I found someone whose commitment to his science hadn't waned, and neither had his personal politics. 'There are simply no races in humankind,' he wrote to me. He remains a hero to biologists in his field, an inspiration, someone who had helped build his scientific discipline into one that today has enormous importance. It is impossible not to admire him. And yet, it is also difficult to read his work and come away convinced

that his generation of scientists had fully abandoned race science after the Second World War. Although they had ditched
race in name, it wasn't clear that they had necessarily shed it in
practice.

In 2000, after almost a decade of controversy around the
Human Genome Diversity Project, and with the project no
closer to being realised, Cavalli-Sforza published a book titled
Genes, Peoples and Languages. In it, he deftly sketched his
grand plan for how genetics could be used to reconstruct human
history. It's also a story, the book's back-page blurb added, that
claimed to reveal 'the sheer unscientific absurdity of racism'.

In a section titled 'Why Classify Things?' Cavalli-Sforza
wrote eloquently about taxonomy, and how humans have always
felt compelled to categorise objects. Yet somehow, he managed
to write this without any reference whatsoever to politics or
social history. He never mentioned that humans were classified
in large part because it was politically and economically useful
to those who did it. He completely glossed over colonialism
and slavery, and the ways in which they fundamentally shaped
how European scientists thought about race in the nineteenth
and early twentieth centuries (indeed, within his own lifetime).
Instead, as he saw it, racism was merely a scientific idea that
turned out to be incorrect. 'It seems wise to me,' his chapter
concluded, 'to abandon any attempt at racial classification
along the traditional lines.'

It's easy to miss the catch. And it lay in the final four words of
his statement: *along the traditional lines*.

'A race is a group of individuals that we can recognize as
biologically different from others,' he continued. Clearly, then,
he hadn't abandoned the use of 'race' at all. Going by this statement, there may be no room for three, four or five old-fashioned
racial types, with hard divisions between them – the way we

usually think about race – but there could certainly be thousands of 'social groups' all over the globe, each characterised by its gene frequencies, and therefore having some biological distinctiveness to them. This is the obvious result of relatedness. People who are related are of course closer to each other genetically, and historically we have tended to live near our relatives, which is how clustering of genetic similarity happens. This means that even neighbouring towns may be genetically different from one another in some slight but statistically significant way. The small clusters produced by the fact that we don't mate completely at random could, by Cavalli-Sforza's definition, be considered 'races'. So according to his own definition, there *are* races, except the number of them is practically endless.

This is an old idea, which owes itself to early geneticists like Cavalli-Sforza himself who wanted to move the study of race away from vague generalisations and make it more precise. As early as the 1930s, when the field of population genetics was just emerging, evolutionary biologist Theodosius Dobzhansky – later an inspiration to Cavalli-Sforza – was the one to substitute the old-fashioned idea that races were fixed types with the more modern idea that they were populations sharing certain gene frequencies. Like Cavalli-Sforza, Dobzhansky was an outspoken anti-racist. But while being actively involved in anti-racist efforts within the scientific community, Dobzhansky also retained the concept of race. He just redefined it. The way it was redefined squared the circle of how it was possible for all humans to be practically the same while also being different. Under this designation, there's no contradiction in my having possibly more in common genetically with my white neighbour than with my Indian one. My population group as a whole (say, north Indians) will share genes in frequencies that her population group (say, white Britons) doesn't. In other words, if you want it to, race can exist, but you must remember that it's statistical. Not every individual will fit.

'They basically redistributed race,' argues Joanna Radin. According to her, the problem with this statistical 'population' approach to studying human difference is that even though it may look different in some ways, it hasn't fully discarded the baggage of the past. 'An interesting analogy would be colonialism,' she tells me. 'A colonial nation declares independence and they have to forge a new nation with the structures of the old colonial regime, and it's very, very hard to transcend that.' Even if the word 'race' isn't being used, the idea of race is still there, deep within the bedrock.

Canadian philosopher Lisa Gannett has similarly warned about the ethical limits of thinking about race in this new way. To some, it may not seem racist to think about average 'populations' rather than distinct 'types' of people. Certainly, early population geneticists such as Dobzhansky believed that racism was rooted in the assumption that within ethnic groups, people are all the same, whereas those like him believed that, within these groups, people are actually very different. But in the racist mind, as Gannett explains, it doesn't necessarily matter how differences are distributed, so long as they are there in some form or another. This conceptual loophole in population genetics – the fact that we're all different as individuals but that there is also some apparent order to this diversity – is what has since been seized upon by people with racist agendas. Gannett calls it 'statistical racism'.

The question all this raises is a slightly odd one: Is race still a problem if we redefine race? And even odder: Can science be racist if the people doing it are anti-racists? For Radin, intention does matter, but it doesn't fix the underlying problem 'If you look at the UNESCO statements on race, people often think that they declared that race is a social construct and that race doesn't exist. But really what they did is try to constrict use of the term "race" to biologists who could be seen to use it responsibly, and not equate it with inequality.' In the hands of liberal, left-wing, anti-racist population geneticists, race was thought to be safe, because their politics were beyond reproach. 'I do think that their sense of virtue really emboldened them to feel like they were creating a transcendent mode of science, that might be able to leave race behind,' she explains.

In *Genes, Peoples and Languages*, Cavalli-Sforza added a humorous aside when talking about the fact that researchers who looked hard enough could spot average genetic distinctions between neighbouring populations, even at the village level: 'People in Pisa and Florence might be pleased that science had validated their ancient mutual distrust by demonstrating their genetic differences,' he quipped. But then, isn't this exactly what racism is? A dislike of others in the belief that they are biologically different? In the mind of the racist, it probably doesn't matter how big the groups happen to be, or if the differences are gradual or sharp. It presumably means equally little if it's all about gene frequencies or population averages, so long as the differences are real. If the people of Pisa and Florence could have their mutual distrust validated by population genetics, then why not the people of any other two places?

For Radin, the problem is obvious. It lies in the need to group in the first place, to separate even when that separation means having to zoom in on the very tiniest bits of the genome that might differ, and even then only on average. This need to separate, to treat people as different, is how race was invented. 'What happens is that you've got a large community of very well-meaning, self-described anti-racist scientists seeking to find a way to move beyond race into population genetics, which seems to be incredibly neutral. It's numbers, it's statistical, it's objective,' she says. 'What they have a more difficult time reckoning with is that even something like population genetics is a science done by people, working with the assumptions and the ideas that are available at the time.' Such scientists may believe themselves to be free of racism, but they can't help thinking about humans in racial terms.

In his correspondence with me, too, Cavalli-Sforza made a comment that betrayed this problem. He observed that interracial or mixed heritage relationships – relationships that in the early twentieth century, eugenicists feared might lead to offspring with strange physical and mental deformities – have turned out to be no bad thing. Miscegenation, as it was called in the past, is obviously no threat to human health. He wrote in his message that you only have to look at 'the beauty and vitality of hybrids, children of partners coming from genetically distant groups' for proof of this. It was his use of the phrases 'hybrid' and 'genetically distant' that disturbed me. This kind of language might have seemed at one time scientifically acceptable, but is it any more? It implies that human populations are like different breeds, even different species.

There is a grey area in which well-meaning people make what in other contexts might be considered incendiary statements, and we overlook them because we know these people are well-meaning. In reference to reproduction rates, Cavalli-Sforza's book stated, 'Europeans are largely at a standstill while populations in many developing countries are exploding; thus blonds and light-skinned people will decline in relative frequency.' If the superficial differences between us don't matter, then why should this? To the population geneticist, that people with blond hair are disappearing may be as much of a concern as the possibility that Native Americans might dwindle, or that Andaman Islanders living in the Bay of Bengal might be subsumed into the wider Indian population. To the anti-racist, objective scientist, there is no value judgement in this. It's a problem to science only because researchers are losing some interesting subjects of study, some statistical corners of human diversity, perhaps a few blue-eyed, blond gene combinations. But to someone with alternative politics, it might be seen instead as an argument in favour of racial purity, of preserving distinct population groups against the threat of miscegenation.

It's easy for academics to imagine that the language they use, and the frameworks they operate in, don't really matter. They are just words, not data. 'I think that in the real world what the scientists say has about as much influence as turning on a fan does on El Niño. It's what throwing a cobble into the English Channel has on Atlantic weather,' Henry Greely tells me near the end of our interview. 'We're just not that important.' But it does matter, because their frameworks and language contribute to our understanding of ourselves. If scientists call people of mixed ancestry 'hybrids', this implies that race is real because we are different enough to warrant using that word. If they talk about 'isolates', this sounds like there are groups who are more 'racially pure' than others. Dismantling the edifice of race is about more than just tweaking language, it is about fundamentally rewriting the way we think about human difference, to resist the urge to group people at all.

It takes some mental acrobatics to be an intellectual racist in the light of the scientific information we have today, but those who want to do it, will. Racists will find validation wherever they can, even if it means working a little harder than usual. And this is the reason that good scientists who do reliable research, ones who are also well-intentioned and anti-racist as Cavalli-Sforza was, can't afford to be cavalier or leave room for misinterpretation. There's an uncertain space between recognising that there is a gap in knowledge and actually filling that gap. It's a place where speculation thrives, where the racists reside. Racists adopted the same concepts as good scientists and the same language as anti-racists to claim that, if some average differences can be seen between certain groups then, by that logic, certain groups *might* be better on average than others at certain things. When Steve Sailer and his followers talk about 'human biodiversity', this is what they mean. This wolf in sheep's clothing is twenty-first century scientific racism.

We were told this might happen. A caustic report on the Human Genome Diversity Project released in 1995 by UNESCO's International Bioethics Committee sounded precisely that warning. It argued that the project could, whether it meant to or not, give racists some basis to believe certain groups were inferior or superior to others. In particular, the committee was concerned that by bringing genetics to the fore in telling the human story, people would ignore culture and history, and return to the kind of simplistic biological thinking that propelled the eugenics movement in the early twentieth century. It advised scientists to resist the temptation to use their work to shore up any kind of political ideology, whether racist or anti-racist. 'Racism,' it reminded them in case they had forgotten, is 'socially and politically constructed.'

Science is just a pawn in the bloody game.

*

Although the work itself never got off the ground, the concept behind the doomed Human Genome Diversity Project did survive. In the years that followed, other teams stepped in to achieve essentially the same outcome in other ways. The Center for the Study of Human Polymorphisms in Paris today keeps a bank of DNA samples from populations all over the world, ready for researchers who want to tap it. In 2002 the United States National Human Genome Research Institute introduced a $100 million initiative to study human variation. And in 2015 the United Kingdom launched its own project to make a genetic map of the people within its own borders, named People of the British Isles.

The project had one more unintended consequence. In 2005 the National Geographic Society in Washington, DC, the one behind the famous magazine and satellite TV channel, decided to dip a toe into the world of population genetics. Naming its effort the 'Genographic Project', it chose anthropologist, geneticist and television presenter Spencer Wells to lead it. Wells had spent a portion of the 1990s studying under Luigi Luca Cavalli-Sforza, seeing the difficulties around the Human Genome Diversity Project at close quarters. His solution to the controversy was simple. National Geographic would sell easy-to-use kits, helping people understand the history of migration that might be hidden in their DNA, and in the process build a data bank from their genetic information.

'We put together this consortium of scientists with the goal of sampling the world's DNA, and at the same time wanted to enable anybody, any member of the general public who was curious about their own genetic ancestry, to get themselves tested,' Wells tells me. The idea that people might want to spit into a cup and have their ancestry tested didn't seem at the time like a highly profitable venture; even the CEO of National Geographic warned Wells before the launch that nobody was going to spend a hundred bucks to test their DNA.

It turned out to be a money-spinner. 'The day we announced, we sold ten thousand. It had hit a hundred thousand by the end of the year,' he says. 'It launched the consumer genomics industry.' The trade was given a noticeable boost in 2006 when media legend Oprah Winfrey had her DNA tested for a television show, revealing ancestral links to people now living in Liberia, Cameroon and Zambia. She also turned out, unlike many black Americans, to share no recent ancestry at all with Europeans. 'I feel more connected to where I've come from,' she told the presenter, Henry Louis Gates Jr, Harvard professor of African American Studies, who wrote a book about Winfrey's experience.

Before long, companies such as 23andMe and AncestryDNA were selling their own kits, turning over billions of dollars. In 2018 it was announced that AncestryDNA alone had sold a total of around ten million kits around the world.

Spencer Wells has since left the Genographic Project, become the owner of a nightclub in Texas, and moved on to new genetic testing ventures. He tells me that the ancestry testing industry flourishes in the United States because of the rootlessness of so many of its citizens. 'In societies like the present-day US, where we have a lot of hyphenated Americans – Irish-Americans, Italian-Americans, Hispanic-Americans – people feel somewhat disconnected from the entity that comes before the hyphen. So they want to figure out who those people were. Who were the people who migrated to the US?' For African Americans in particular, many of whom were transported as slaves and ripped away from any connection to their families or homelands, the kits offer the only means they may have of tracing their genealogy. The psychological effect on the public of sequencing the human genome, of convincing them that our differences are identifiable in our DNA, is that this now appears to be a foolproof way to define who we are.

In reality, genetic testing is only an educated guess about where your relatives may have lived based on the data fed into the models in the first place. As geneticist Mark Jobling explained, it's possible to group people any way we like. Ancestry tests scan portions of people's genomes to find those who have genetically a little more in common, then pool them together. Theoretically, they could pool them using any measure, even as simple as those who live south of the equator and those who live north of it. But of course, companies most often use old-fashioned racial categories or nationalities. This means that if there are few or no DNA samples from people in the country your ancestors came from, you're stuck. One of the reasons Oprah is linked to Liberia, for example, may be that this is where former slaves were long ago 'repatriated' by white American leaders who couldn't bear the thought of such people living freely among them. Ancestry testing doesn't show you your past as much as it reveals the people you are distantly related to in the present, and even then only if they have had similar tests done. Oprah has some connection to people who now live in Liberia, but this is not necessarily her ancestral homeland, the place from which her relatives originally came.

Mark Thomas at University College London, a leading geneticist who has seen his own research recruited into these models, tells me he has always been sceptical of ancestry testing firms. 'It's not that they're cynical, it's not that they're nasty, it's not that they've got particularly racist agendas. They want to make money, and you make money by servicing people's prejudices.' Not just prejudices, but also people's natural desire to know *who they are*. In 2018 Thomas and his colleagues published a paper detailing how one particular firm based in Scotland, BritainsDNA, had threatened them with legal action after they challenged the firm's wide-ranging claims in the press – for instance that the actor Tom Conti is 'Saracen' in origin, and that Prince William has Indian ancestry. They didn't have the data to establish either claim.

What ancestry testing has done is take the work of well-meaning scientists, who only tried to do good in the world, and inadvertently helped reinforce the idea that race is real. Using their methods and data, an entire industry has achieved exactly the opposite of what scientists like Cavalli-Sforza once set out to do. The true way that variation works, the nuances, are rarely explained by those selling ancestry testing kits. Having seen how purely 'black' Oprah Winfrey really was, for example, white supremacists in the United States began using the very same tests to prove how 'white' they were, sometimes sending off vials of spit to various companies until one came back with the desired result, establishing beyond doubt that they were of nothing but European ancestry. By forcing people into categories, even if that means dividing our individual bodies into so much European, so much African, so much Asian and so on, the tests fortify the assumption that race is biologically meaningful. If it's possible to categorise, we assume, there must be something to the categories.

The irony is that as more research has been done into our origins since the launch of the Human Genome Project and consumer ancestry testing, it has only undermined these measures of identity. Within the last decade, as scientists have uncovered exponentially more genetic evidence about us and our ancestors, even they have been surprised by the results. Nothing has matched expectation. Our roots, it turns out, are very rarely where we think they are.

7

Roots

What race means now in the
light of new scientific research

WHEN SHE WAS GROWING UP, my little sister was a diehard fan of Morrissey, frontman of The Smiths, genius songwriter, British cultural icon. For one of a handful of brown girls in a white working-class south-east London suburb, indie music spoke to that cold, lonely feeling of not quite being able to fit in. If the British National Party was marching outside our door, inside her headphones was a different British voice that she could relate to. He was a refuge from those who insisted that we all had to be the same.

But in an interview with a music magazine in 2007, Morrissey said something that couldn't help but trouble my sister, as well as other fans. 'Whatever England is now, it's not what it was and it's lamentable that we've lost so much,' he complained. He railed against high immigration, against what he saw as a change in the character of Britain. There was public outrage. She lost a hero. But as we in our family knew too well, out in the country as a whole there were many who felt this way. This was a debate that had been simmering for decades, occasionally stoked by national politics, making people anxious, wondering what it meant to be British.

A decade later, the pot bubbled over. A financial crisis and economic austerity, coupled with higher than usual rates of immigration from eastern Europe, helped fuel support for nationalists who wanted to cut the country free from the continent. In a referendum in 2016, the majority of voters agreed that leaving the European Union might be a good plan. They were promised a new dawn. The nation would stand alone, the way it had done during the days of Empire, riding the waves of unbridled trade and setting its own rules on who would be allowed into the country.

For visible immigrants, or children of immigrants like my sisters and I, watching this could sometimes feel like an out-of-body experience. The borough in which my parents lived, and where we grew up, was one of only five out of the thirty-two in London that voted to leave. As citizens, we had the right to vote to decide Britain's future, but we also knew that a sizeable slice of other voters wanted fewer of us there in the first place. A campaign poster showed legions of men with skin as brown as ours, queuing up against the slogan 'Breaking Point'. The far right was emboldened. Around the time of the referendum, reports of race-based crime rose, with a sharp spike in the kind of everyday racism that I last saw as a teenager.

Squaring your appearance with your nationality is one of the hardest things about being an ethnic minority. Not all, but some of those who voted to leave Europe wanted a return to their own particular vision of Britain. Skin colour mattered to them because it was a visible baseline, the reference. It was the way the British had always looked, from the beginning, before Empire, before Shakespeare, before kings and queens, before culture and values. Britain, as far as we were aware, has been forever white. So even though I was born in London, I speak the Queen's English, my dinners generally comprise meat and two veg (I probably eat a curry as often as most white Britons, and possibly less than some), and my wireless is set permanently to Radio 4, in their eyes it is my failure to be the right colour that truly undercuts my claim to Britishness.

What nobody could have predicted then was that, by an almost cosmic coincidence, at the very moment Britons were struggling to define their identity in the face of political turmoil, and particularly for those racists who saw Britain as a white nation first and foremost, some news was coming. They were about to be thrown a curve ball.

*

I saw it for myself at London's Natural History Museum in 2018, a package no fancier than a bunch of old bones.

The skeleton is laid out neatly in a small corner of the museum. Most of the visitors don't linger as I do. To be honest, it looks unremarkable. But this is the frame of one of the oldest dead bodies ever found in the country, dating from some 10,000 years ago. And it's full of secrets. Almost as soon as he was discovered in caves in Cheddar Gorge in Somerset in 1903, earning the name 'Cheddar Man', people began to wonder how he must have looked. They wanted to put a face to one of our early ancestors. Archaeologists could certainly guess that he was short by modern standards, that he probably had a good diet, and he may have been around twenty years old when he died. One speculative reconstruction showed him to be white-skinned, with rosy cheeks and a trailing brown moustache. But his actual appearance was a mystery.

The genomes of living people offer a limited and fuzzy picture of the past because of mixing and migration. We're just so similar. This is where the study of the bones of our distant ancestors and their ancient DNA has come in. It has achieved what the Human Genome Diversity Project couldn't. When it comes to tracking human migration patterns over thousands of years, even archaeology and linguistics can't provide the same historical data that ancient DNA can. By around 2010, genetic sequencing techniques had developed far enough to tease out highly reliable samples of DNA from ancient specimens (a bone just behind the ear turned out to be best) and use them to help reconstruct entire genomes of long-dead people. The use of this technique has mushroomed in the last decade. It has been credited with solving historical mysteries at a stroke. Thousands of skeletons from all over the world have been analysed already, and as the British public were about to learn in early 2018, Cheddar Man would be among them.

Scientists at the Natural History Museum and University College London revealed that he probably had blue eyes and curly hair – no great surprise here. But what came as a real shock to many was that his bones also carried genetic signatures of skin pigmentation more commonly found in sub-Saharan Africa. It was probable, then, that Cheddar Man would have had dark skin. So dark, in fact, that by today's standards he would be considered black. The revelation, along with a dramatic new reconstruction of his face markedly different from the original one, made front-page news and television bulletins.

'Hard cheese for the racist morons,' smirked a headline in the *Mirror*. 'Another racial panic for white supremacists,' announced the news website *Salon*.

Panic was indeed sparked. People experienced all the stages of grief. On far-right websites, a few immediately began

doubting the scientific results – maybe, just maybe, the researchers had got it wrong. Some hopefully voiced the possibility that
Cheddar Man hadn't been an actual Briton at all, but was just
a passing visitor who happened to die here, like an unlucky
tourist. Finally, there was acceptance. Some, especially those
who for so long had believed that skin colour was the basic
measure of Britishness, wondered if perhaps it was time to
rethink national identity. If the original Britons were black, all
bets were off.

*

Throughout the frenzy, there was one set of people for whom
the news barely registered a flicker on their excitement dial.
They weren't surprised at all.

'With the whole Cheddar Man thing, I was amazed initially at just how much press coverage it got,' I'm told by Mark
Thomas, professor of evolutionary genetics at University College London, who worked on the finding. Leaning back in his
chair, wearing stonewashed jeans and a grandad-collar shirt,
Thomas is about as relatable as a professor comes. He is one
of the world's leading experts on ancient DNA and, from this
position of authority, he has a tendency to tell it how it is. For
geneticists like him, the Cheddar Man discovery was unremarkable given what they already knew. They had more or less
expected it. 'What I was even more amazed about was the *Daily
Mail*-reading backlash,' he laughs.

Thomas had welcomed the outcome of the Cheddar Man
tests as just another piece of evidence in a huge body of research.
It took up a couple of sentences in his latest paper. Scientists had
already known for a few years, from analysing the skeletons of
other hunter-gatherer bones found in western Europe, that dark
skin pigmentation could well have been common back then.

After all, light skin was likely an evolutionary adaptation, one that helped people living in northern climates absorb more vitamin D when there wasn't enough sunshine. The first human pioneers didn't arrive in Europe or Asia looking white, because they had originally migrated from Africa, where there was little or no survival advantage in having lighter skin pigmentation.

What researchers were a little less sure about was how quickly paler skin emerged, where and when. 'Over the last 10,000 years? Or over the last 40,000 years?' asks Thomas. One theory was that it developed very slowly and gradually, starting 40,000 years ago when modern humans first came to live in Europe. Another suggested it was a more recent phenomenon, contemporary perhaps with the advent of farming. Trading a hunter-gatherer lifestyle for settled agriculture would have limited people's diets and made it even more vital that they get the vitamin D they needed from the action of sunlight on their skin. Another is that light skin emerged elsewhere in the world, outside western Europe, and that the movement of people would have then introduced it to darker-skinned Europeans. Evidence as it stands indicates that, like Cheddar Man, many other pre-farming hunter-gatherers who lived in western Europe during his time and at least as recently as 7,000 years ago would have had light eyes, dark hair and dark skin. It was the first farmers to come into the region later from the east who brought with them their lighter skin and brown eyes.

So one thing was clear: Cheddar Man wasn't an exception in his time. People all over the world then didn't look anything like the way we look now. Not only this, they looked more different from each other than we do today.

'Differentiation between groups in different parts of the world would have been greater,' explains Thomas. The scientific explanation for this is genetic drift. Being in small groups as they were, every breakaway bunch of migrants as it moved began to look more and more different from the relatives they left behind as time passed. Since then, as groups have grown bigger and remixed with each other, populations across the world have become more homogenised. Today it's generally possible to know whether someone's ancestry may be rooted in, say, Asia or Europe or Africa by looking at them. But ten millennia ago, we would have struggled to do this. Appearance didn't map the way it does now, and physical features in some regions may have been dramatically different.

When Thomas and his team looked into the very earliest farmers in the Fertile Crescent, who lived in what is now Iran, and compared them with farmers in nearby Anatolia and the Aegean, they found to their surprise that the two were genetically very distinct from each other. 'They were as different as people from Ireland and Thailand today, more or less. I mean, of that order of magnitude.' Today, neighbouring populations tend to be much more similar. They've mingled and mated with each other, mostly dissolving away the gaps.

Our modern ideas of race are deeply connected with how we look. Our appearance is a shorthand for the stereotypes, a means of slotting people into groups and making judgements about them. The disbelief that met Cheddar Man's probable blackness arose because many among the British public couldn't help but assume that Britons had always looked a certain way, even in the distant past. They struggled to categorise him, forgetting that he existed thousands of years before our racial categories came about. He was proof that there couldn't be anything eternal or pure about race because once upon a time, not so very long ago in evolutionary terms, most of the people on earth didn't look like us. They were already human. They were, however distantly, *us*. But they looked different. The long lens of evolutionary history has a way of turning all you think you know on its head.

The picture becomes even more complex as we go further back in time. Geneticist Sarah Tishkoff at the University of Pennsylvania has carried out pioneering studies into skin colour variation across Africa, finding that the genetic variants – different forms of the same genes – associated with both dark and light skin have existed in Africa for a long time. The variants associated with light skin are common not only in Europe and sometimes in east Asia, but also in the San hunter-gatherers. 'These are the people in southern Africa who have the oldest genetic lineages in the world,' she says. This suggests that rather than evolving independently outside Africa, many of the gene variants associated with light skin may already have been there when people first migrated from the continent.

So not only did people with darker-pigmented skin occupy Europe, but even earlier, there were genetic variants for lighter-pigmented skin in Africa. Given the evidence so far, Tishkoff suggests that lighter-pigmented skin may even have been the ancestral state in the long-distant past. Underneath their bodies, chimpanzees – our closest genetic relatives – tend to be light skinned, their dark body hair providing protection against the sun. 'When our ancestors left the forest and went to the savannah, there would have been selection for better thermo-regulation, so getting rid of body hair, increasing the number of sweat glands. And if you're decreasing the body hair, there would be selection for darker skin.' Darker skin could have been one of the adaptations to a new living environment within Africa.

Yet when scientific reconstructions are made of earlier human species, such as *Homo erectus*, they are almost always given dark skin. There's an assumption that our species began as black. 'I don't think that's necessarily the case, because both light and dark variants have been around for a really long time. And there could have been variation in Africa a million, two million years ago,' explains Tishkoff.

Even today, there is far more variation in Africa than the simplistic black–white model of race implies. 'I think many people don't recognise the large range in skin colour in Africa . . . The whole continent of sub-Saharan Africa is incredibly diverse genetically. It doesn't fit with a racial model, one homogeneous African race. There's a huge amount of variation amongst populations in Africa,' she adds. 'Skin colour is a terrible racial classifier. There really are no good biological classifiers for race.'

For the biologists who know this, skin colour begins to lose its meaning. 'I mean, it's skin pigmentation, you know! It's just so trivial,' reflects Mark Thomas. He found reaction to the new finding about Cheddar Man bizarre given the scientific facts. 'Obviously there are some idiot racists over there in the corner for whom it is important. But I think that if you base your identity on the pigmentation of some West Country bloke from 10,000 years ago then you really should rethink it. My own personal view is that today we over-privilege and fetishise the concept of identity.'

Thomas reminds me that the physical features we associate with race are poor proxies for overall genetic similarity. Even if one population tends to have darker skin and another lighter, that doesn't mean their genomes as a whole will have less in common than two populations with the same skin colour. Variations in physical appearance, whether it be skin pigmentation, ear shape, nose shape, whatever, says Thomas, makes the gaps between groups feel far larger than they really are genetically. In biological terms, the differences really do appear to be no more than skin-deep. It's an error to assume that the internal

differences are as profound as the external ones appear. But it's
an easy one to make. 'If we could see each other by looking
at our genomes then, without a big computer, you would be
hard pushed to work out whether somebody was from India or
from Poland,' he explains. One might assume, for instance, that
light-skinned Irish people would have very little in common
genetically with darker-skinned south Indians. But that's not the
case. 'There is relatively little genetic differentiation between
southern India and Ireland. I mean, relatively similar ancestry
components. But of course, the pigmentation differences are
quite large, and so people assume that these people are mas-
sively different genetically.'

In that sense, how we look is misleading. 'Nature plays dirty
tricks on us,' says Thomas.

It can play tricks on scientists, too. If data seems to suggest
that populations are very different, for the most part it's because
population geneticists are deliberately examining the small sections of our largely shared DNA that happen to differ. This is
their job. 'We're zooming in. We're turning up the contrast on
what are actually tiny little differences over extremely closely
related populations,' he warns.

*

'The past is very surprising,' says David Reich, a geneticist in
the ancient DNA laboratory at Harvard University. 'It's different
from how most people picture the past in their heads.'

Reich is the most well-known person in this branch of
science, at the forefront of using genetics to plot ancient migrations around the world. At the moment I happen to visit him,
though, he has become embroiled in controversy for suggesting
in the press that more work needs to be done to understand
cognitive and psychological differences between population

groups, a phrase that most people have interpreted as meaning 'racial differences'. His statement – a departure from the nearly seventy-year consensus that studying race isn't the business of biologists – has attracted angry emails from fellow academics. But he hasn't backed down. When I see him, I expect him to be defensive, maybe even brash.

I couldn't be more wrong. With his hands in his lap, so softly spoken that my voice recorder struggles to pick up every word, he surprises me with his gentleness. His half of his office is bare, save for a few drawings stuck to the plain white walls. He is unfailingly polite, pausing only to message his wife. The one clue to his global importance is the steady stream of students and researchers lining up to see him outside. One young man sits with his laptop at a bench in the corridor all day in the hope Reich may be able to spare him a minute or two later on.

Reich's lab is a powerhouse. It has scoured the world for skeletons that might provide genetic evidence of the past, and as Reich has noted, it churns out findings so quickly that the amount of data doubles faster than the time it takes for it to be published. Scientific journals simply can't keep up. But for him, this is more than a scientific gold rush. Genetics has a way of cutting through ancient historical questions in a way that nothing else can. His group, along with the lab of Mark Thomas and others across the world, helped confirm the hypothesis that farming emerged 10,000 years ago in the Near East – the region between Europe, Africa and Asia once spanned by the Ottoman Empire, and before that, by Mesopotamia – and that these farmers may have been a genetically varied group of humans who then helped spread agriculture to other parts. He is also fairly confident that natural selection has caused southern Europeans to be a little shorter on average than northern Europeans.

But it's the story of migration that is the most revealing. What we think of as 'indigenous' Europeans are, scientists like Reich now understand, the product of a number of migrations across the last 15,000 years, including from what is now the Middle East.

The British, in particular, have their own story to tell. 'Britons in the past didn't look like Britons today, and were genetically very unlike Britons today,' Reich explains. Whoever the first inhabitants of Britain were, their way of life was likely to have been replaced almost wholesale between around 5,000 and 4,000 years ago by a group of people coming through Europe from the steppe grassland that stretches between the Black Sea and the Caspian Sea. They are known by some anthropologists as 'Beaker folk' for their distinctive bell-shaped pottery. While artefacts of Beaker culture are found scattered all over Europe, the team here at Harvard has shown by studying the DNA of four hundred ancient Europeans that these people must have swept in and supplanted almost everyone who was living in Britain at the time as well.

How they did such a thing is unclear. They could have simply come in large numbers and bred with people who were already there. They may have been better equipped to survive in the environment, through resistance to certain diseases or by virtue of their technology. The pre-existing populations could have been collapsing already, as some data suggests. Whatever the explanation, their arrival changed not only the culture but also the way people looked. The steppe people with their Beaker culture had lighter skin. According to estimates drawn up recently by Reich and his colleagues, this Beaker invasion replaced around 90 per cent of Britain's gene pool. And all in the space of just a few centuries.

This means that light skin did not define Britons from the beginning. 'There's been a continuous process of skin lightening with big jumps that occurred at these migrations,' Reich suggests. 'So for example, when the first farmers came to Britain about 6,000 years ago there is a big change in the average hue of skin around then, predicted by the genetics. And then when this Beaker phenomenon spread into Britain, there was another big jump associated with that.'

While some of this confirms what archaeologists already suspected, it's more surprising just how much churn there has been in global patterns of migration throughout the ages. Reich himself was taught when he was younger that humans spread out from Africa, with little mixing once they started to split, like branches of a tree. Once they landed somewhere, people stayed put. That was the common assumption. But the evidence that's now emerging suggests something entirely different. 'It became very clear that the big large-scale mixture, migration, or gene flow, however you want to call it, is common and recurrent.'

The true human story, then, appears to be not of pure races rooted in one place for tens of thousands of years, but of constant mixing, with migration both one way and another. The cherished belief that people in certain places have looked the same way for millennia has to give way to the understanding that migration made the world a melting pot long before the last few centuries, long before the multicultural societies we have today. Our roots are not an orderly family tree but instead are tangled, according to Reich, more like a climbing plant on a trellis. Our ancestors branched out but then came back, and remixed, again and again throughout the past.

'I think this idea of indigeneity, and you being from a population that has been here for ages – I mean there may be populations that have better claims to that than others – but at some deep level the great majority of people in the world, if not everyone, is not derived directly from people who lived in the same place deep in the past.'

The British story is just one of thousands. For example, the Beaker folk were part of a far earlier, bigger and longer migration out of central Eurasia and into many different corners of the world of people associated with what archaeologists call the Yamnaya culture. They were pastoralists, raising and moving livestock, with wagons and horses that made them mobile in a way that may never have been seen before. Their diet was rich in meat and dairy. From roughly 7,000 years ago, across a span of two millennia, the Yamnaya (themselves a product of earlier migrations into the region they came from) trekked west and south-east, not only populating Europe but also going as far as north India. They brought the wheel and, it has been mooted, they also brought cannabis.

By 5,000 years ago, the Neolithic farming cultures of Europe had been largely replaced. Kumarasamy Thangaraj at the Centre for Cellular and Molecular Biology in Hyderabad tells me that around the same time, those of the Yamnaya culture came in from the north of India and mixed with the people who were already there. The Indian population was itself a mix of indigenous hunter-gatherers who had originally moved out of Africa many thousands of years earlier, and more recent farmers migrating from what is now Iran. All Indians, save the tiny community of Andaman Islanders who have been closed off from the Indian mainland for thousands of years, are a blend of these three ancestral populations.

Confirmed by genetics, these ancient connections can be spotted in the words we use. Linguists long ago saw similarities between European and Indian tongues, describing them together as Indo-European languages. Genetics has added more hard data to the history. Almost all Indians today are genetically connected to Europeans by their ancient ancestors who spread the Yamnaya culture, as well as the earlier farmers migrating from the Middle East.

*

'If you pay any attention to the discoveries coming out of science, they don't play into any sort of old systems of prejudice,' David Reich tells me.

Take Stonehenge, the mysterious prehistoric assembly of standing stones in south-west England, which attracts more than a million visitors every year. Within a few hundred years of its construction around 5,000 years ago, the Neolithic farmers who constructed it were largely gone. They were probably replaced by incoming folk who followed the Beaker culture, because within another thousand years, Reich's team could see little evidence of Neolithic ancestry in the genomes of ancient remains they were studying. Now, just pause to think about what this means: the symbol we associate with ancient Britain, the one thing that couldn't really be more authentically British, was built by people who are certainly not the main ancestors of those who consider themselves indigenous Brits today.

Cheddar Man and his relatives, too, who lived 10,000 years ago, couldn't have been from the same genetic population as Britons today, because like the builders of Stonehenge, they were replaced by farmers who spread across Europe from Anatolia. Cheddar Man and his people don't have any direct descendants – only bits of them exist now, explains Mark Thomas. What he means by 'bits' is that Cheddar Man and his relatives on the continent would have bred with whoever came into the region. So while his own particular population and their culture didn't survive intact, traces of them would have endured, either because they mixed to some extent with farmers coming into Britain, or because their continental relatives mixed with farmers spreading across Europe.

To know that this melting pot has been churning for thousands of years puts a fresh spin on the contemporary idea of race. 'I think that genetics and genomics have a wonderful opportunity to undermine these outdated and scientifically unsupported notions of race, ancestry, ethnicity and identity,' says Thomas. The feeling that there is a 'home' for us all, and that our bodies somehow reflect this, deeply and viscerally, begins to melt. The attachments we have to places and their relics, the ancient stories that tell us who 'our people' were, have to be rethought when we understand that 'our people' were actually migrants into a place occupied by others. The relics belong to them, not us. The place was theirs once, before others came along. We're all part of the churn.

What is even more mind-bending is that when you're looking this far back in time, ancestry expands to include almost everyone. 'Cheddar Man's people are to an extent the ancestors of just about everybody in Europe,' he explains. 'Indeed, it is possible that in his group are the ancestors of everybody in the world just about, maybe everybody in the world today.'

This may seem implausible, but it's just mathematics. The further back you go in time, the weaker your genetic link to your ancestors. Five generations ago, you would have as many as 32 possible ancestors contributing to your genetic makeup. Nine generations back, you could have 512, many of whom may have contributed next to nothing. Zoom back fifteen generations – still just the tiniest slice in recent human history – and you could have 32,768, assuming nobody was having babies with someone they were even distantly related to, which is unlikely.

Each of these could give you no more than the tiniest fraction of your DNA. Longer and longer ago, the theoretical number rises into the millions, and ultimately, to more people than were alive at the time. Of course, that is impossible, the only explanation being that we are all at least a little inbred.

Even if you could trace your lineage as far back as Cheddar Man, or more recently to Nebuchadnezzar or Cleopatra or any other figure from antiquity, you would probably be no more related to them than is a random person on the street. The more you zoom into the past, the more your ancestral history begins to overlap with that of everyone else on the planet. As Thomas notes, we only have to go back a few thousand years before we reach somebody who is the ancestor of everybody alive today. Go back a few thousand years more, and everybody who was alive is either the ancestor of everybody alive today (if they had descendants who survived), or nobody alive today (if they didn't). Hence Cheddar Man, if he had children and they had children, and so on until today, is both your relative and mine.

Race, nationality and ethnicity are not what we imagine them to be when seen from the deep past. They are ephemeral, real only in as much as we have made them real by living in the cultures we do, with the politics we have. David Reich tells me that he draws a sense of global kinship from his work on genetics. 'I have a personal way in which genetics is meaningful to me which doesn't involve my own ancestry,' he says, quietly. 'I think that one way of relating to the findings about genetics is that we're all related to each other, and we are all part of a broadly closely related group of people over the last couple of hundred thousand years, with a lot of complexity, and with a lot of mixtures and migrations and reticulations. And we're all part of that.'

But then his tone changes. Even after everything he's said, he doesn't dismiss the idea of race altogether.

٭

David Reich is not a racist. But then neither does he adopt the staunch anti-racist position of the old-school population geneticists such as Luca Cavalli-Sforza, who bravely debated with the scientific racists of their time, wearing their politics on their sleeves. Reich respects Cavalli-Sforza, even writing about how much he has been an inspiration to him. But he confesses that he sees himself as apolitical.

The genetics of human variation are complicated and subtle, he tells me. And his own position on race is similarly subtle. Despite his research revealing the extent of interconnectedness between humans, the great uniting trellis of ancient migration, Reich still suspects there's something worth investigating about group difference. And he leaves open the possibility that this difference correlates with existing racial categories – categories that many academics would say were socially constructed, and not based in biology at all, except in very unreliable ways, such as along crude skin colour lines. 'There are real ancestry differences across populations that correlate to the social constructions we have,' he tells me firmly. 'We have to deal with that.'

He admits that some categories make no biological sense. For instance, 'Latino', as anyone from South America is referred to in the United States. 'Latinos is a crazy category that encompasses groups with different ancestry mixes ranging from Puerto Ricans, who have very little Native American ancestry, mostly African, a little European, to Mexicans, who have very little African ancestry and mostly Native American, European . . . It's a crazy category.' At the same time he thinks that some categories may have more biological meaning to them. African Americans are mostly West African in ancestry and white Americans tend to be European, both correlating to genuine population groups that were once separated at least partially by 70,000 years of human history. When it comes to West Africans and Europeans, he continues, 'there's a long time separating these two groups. Enough time for evolution to accumulate differences. We don't know very much about what those differences are because we're still at the beginning of collectively trying to identify biologically what differences do.'

Reich suggests that there may be more than superficial average differences between black and white Americans, possibly even cognitive and psychological ones, because before they arrived in the United States, each of these population groups had 70,000 years apart to adapt to its own different environment. He implies that natural selection may have acted on them differently within this timescale to produce changes that go further than skin deep. Reich adds, judiciously, that he doesn't think these differences will be large – only a fraction as big as the variation between individuals, just as biologist Richard Lewontin estimated in 1972. But he doesn't expect them to be non-existent either.

They are words I never expected to hear from a mainstream, respected geneticist. Reich is of course not a racist. Indeed, like Cavalli-Sforza, he believes that if race research is done, it will only further demolish racial prejudices. Scientists are concerned with fact, not fiction, his argument goes, and the facts we have so far are simply not in the racists' favour. The more good work that is done, the more it demolishes long-standing racial stereotypes, so by his logic there should be no barriers to doing yet more research, even if it feels risky. 'My feeling about this field has been that, broadly, it makes telling falsehoods more difficult. That's my feeling. It may be self-serving, but that's my feeling. And so I think these surprises, such as ancient Britons were very much more dark skinned than present people . . . I think this is broadly a force for combating prejudice, because it doesn't conform to anybody's pictures they had before.'

While he sees the racists as factually wrong, he also sees some anti-racists – those who insist that we are all exactly the same underneath – as not having the full facts either. 'It's a little bit painful to see very well-meaning people saying things that are contradicted by the science, because we want well-meaning people to say things that are correct,' he tells me. 'The way I see what's going on in this world right now, there are racist people that are just perpetrating falsehoods, and just representing the science in incorrect ways, tendentious ways in order to achieve certain goals. And then there's people whose perspective on the world I agree with who are actually saying things that are technically incorrect.'

Reich himself is technically correct that there *could* be more profound genetic differences between population groups than we are aware of at the moment. But to date, no scientific research has been able to show any average genetic differences between population groups that go further than the superficial and are linked to hard survival, such as skin colour or those that prevent a geographically linked disease. There is no gene or variant of any gene that has been found to exist in everyone of one 'race' and not in another. In London, Mark Thomas, who has collaborated with Reich, remains dismissive of the idea that race is useful to genetics. 'Most researchers, including geneticists, agree that "race" is a socially constructed category . . . There is no categorical imperative in biology, and no need or value in placing people in biological boxes. There are subtle genetic correlations with geographic origin, and physical traits, as well as medical ones, and understanding those correlations is important. But there are no hard borders, just gentle gradients,' he tells me. 'Unfortunately, that doesn't stop people "racial-ising" others, and perhaps that reflects our desire to categorise. Most categories are nonsense, although some may be useful. "Race" is useless, pernicious nonsense.'

The question of whether or not biological research into racial difference is useful still divides the scientific community. What seems to bother Reich above all is that the research simply isn't being done, at least not properly and not enough, so we just don't know how useful it might be. Part of the reason for this, of course, is the longstanding scientific taboo which has kept race off the table in mainstream genetics since the end of the Second World War – although certainly not in social science, which has built an enormous body of work on the topic. We have plenty of data on racial gaps in, for example, income, health and schooling in the UK and the United States. This is because race has been accepted by academics as a social reality, not a biological one. Race affects how we live, but not who we are genetically. Reich, however, appears to find this is unfair. 'We've been silenced by the great anxiety that we feel talking about these things, and by the history of abuse of genetics by people seventy years ago, or eighty years ago,' he says.

Reich is probably not the only scientist who would like to be free of 'the great anxiety' caused by eugenicists and scientific racists. But that freedom would have to come with responsibility. As the devastating mistakes of the nineteenth and twentieth centuries proved, race research never goes well when society is racist. And although Reich insists that biological data as it stands makes racism impossible, I'm not so sure.

Two days after I visit David Reich in his laboratory at Harvard, a party is held at the Cold Spring Harbor Laboratory, the world-class research institution on Long Island that was once the site of the United States Eugenics Record Office. The celebration marks the ninetieth birthday of James Watson, one of the legends of twentieth-century genetics, who helped discover the double helix structure of DNA alongside Rosalind Franklin, Francis Crick and Maurice Wilkins. In 1962 Watson, Crick and Wilkins were awarded the Nobel Prize. Watson went on to become the laboratory's director in 1968, and was crucial in helping to get it funding over the years and building its reputation. A Grammy award-winning pianist is invited to give a performance at the party, with no fewer than eight Nobel laureates among the four hundred guests.

Yet, for years, Watson has been known to hold racist and sexist views. He was famously derogatory about his former colleague Franklin, who did much of the experimental work that helped him make the discovery that led to his joint Nobel Prize. He told the *Sunday Times* newspaper in 2007 that he was 'inherently gloomy about the prospect of Africa', because 'all our social policies are based on the fact that their intelligence is the same as ours – whereas all the testing says not really.'

In 2010 David Reich witnessed Watson's racism first-hand at the Cold Spring Harbor Laboratory when they were both at a workshop on genetics and human history. Watson sidled up to him and asked him something along the lines of: 'When are you Jews going to figure out why you guys are so smart?' Reich was appalled. Watson openly compared Jewish people to Brahmins, high-caste Indians, who are known for being over-represented in universities and high-status jobs. Traditionally, they are India's educated, priestly class. Watson suggested that racial purity combined with millennia of selecting for scholarliness was the key to both Jewish and Brahmin success. He went on to make other racial slurs, about Indians being servile, a trait that suited British colonisers, and about the Chinese, whom he thought had been left genetically conformist by their society.

I wonder what Reich took away from this encounter. If understanding the scientific facts makes it so impossibly difficult to be racist, how does James Watson manage it?

Reich hesitates. 'Well, Watson is, you know, probably more sexist than he is racist,' he says awkwardly. 'I don't know. I don't know. He's like uncontrollable. It's impossible to control Jim Watson. He purposely wants to create, to annoy people, to scandalise people, so I don't know. You can't control everybody. I do think that. So, yeah, I don't know.'

There is a long pause, an uncomfortable half-shrug. 'I just don't know.'

8

Origin Stories

Why the scientific facts don't always matter

THE PAST IS PART OF THE PROBLEM. Not just the politics of the past, of the nineteenth-century theories of race that permeate our subconscious, the colonial hierarchy of races. But the deeper past. The problem is with how we build our ideas of who we are. When biologists try to understand ancient human migration, when they pick through our genomes and those of our distant ancestors, they are participating in age-old efforts to piece together our origin stories.

In China, it's believed that taming the flooding of the Yellow River many thousands of years ago, by a man named Gun and his son Yu, marked the dawn of Chinese agricultural civilisation. It's a legend that helped build national identity, serving a unifying purpose, lending a sense of superiority. Over the centuries, myths take on a life of their own, each generation recasting them to suit their needs until we can no longer tell the difference between myth and history. Before we know it, the glorious tales of our ancestors become our historical facts. Their ghosts become our icons. And of course we need to believe that our

forebears were better than they really were, that they were nothing less than superhuman. The founding myth of Rome is of the
abandoned baby twins Romulus and Remus, suckled by a she-
wolf and rescued by a god. German nationalists told of a blond,
rugged hero dubbed Hermann, who defeated the Romans and
united Germany's disparate tribes. These figures have become
woven into national identity, pulling people together in the
belief in the cosmic power of their founders, cementing their
particular claim on human civilisation.

So science is not enough. We need stories to assert our identity, even if they're held together with the tiniest grains of truth.
There was indeed a German tribal chief named Arminius (from
whom the legend of Hermann is constructed) who spearheaded
a victory against the Romans two thousand years ago in what is
known as the Battle of Teutoburg Forest. And in 2016 Chinese
researchers confirmed that there really had been a giant flood
around 1920 BCE. But legends must have been written around
these facts until the people hearing them could no longer separate the two. The bloody realities became whitewashed over
time, each iteration making the story cleaner, brighter and
more dramatic than the original. And this was necessary, not
just for the sake of gripping narrative, but also because it's tough
to build national pride and a sense of superiority around a dirty
history.

Anthropologist Jennifer Raff, based at the University of
Kansas – in her words a 'middle-class white girl' – grew up on
a powerful origin story of her own. 'Those of us in the United
States have been taught this idea of American exceptionalism,' she says, 'that our country is the greatest country, and is
founded on these wonderful beliefs, this freedom and equality
and democracy.' It's a narrative that rests on the assumption that
European pioneers in the seventeenth century filled a largely

empty land with visions of a better society, deploying their unmatched skill and hard work to cultivate it. As in Australia, the indigenous inhabitants were framed as a dying race. If not gone, then definitely on their way out.

The subtext is that, without white Europeans, civilisation couldn't have flourished in North America. The United States was *theirs* to make.

It's not easy to square this popular founding myth with the more brutal historical facts. Of course, Native Americans weren't primitive, dwindling or scarce as the settlers liked to portray them. When the land turned out to be less vacant than they hoped, European colonisers made every effort to empty it. Thousands died on the 'Trail of Tears' – the forced relocation, following legislation passed in 1830, of several Native American tribes from their ancestral homelands in the south-eastern United States to designated Indian territories west of the Mississippi. Many more were killed by diseases brought by the migrants, to which Native Americans had no resistance. Deaths frequently went unrecorded, which means today we have only the vaguest estimates of how densely populated America really had been before. Genetic analysis published in 2011 suggests that the number of female Native Americans may have shrunk by half upon contact with Europeans five centuries ago.

We know this now. The founding myth becomes harder and harder to maintain. And yet it has strange ways of reasserting itself, even within academic circles, as Raff has found. Her work, trying to understand the distant past and the effects of race and migration, has shown her how easy it is for people, including respected scholars, to resist abandoning popular myths and racialised views of history even in the face of undeniable evidence. Indeed the myth of American exceptionalism is so pervasive that an entire scientific theory exists to explain it, weaving in archaeology and anthropology with the notion that Europeans are the ultimate bearers of human progress. It's known as the Solutrean hypothesis.

Crafted in earnest in the 1970s, the Solutrean hypothesis takes its name from archaeological evidence of certain tool-making techniques belonging to the Solutrean culture, which existed in parts of what is now France and Spain between 23,000 and 18,000 years ago. The Solutrean method of making blades by forming long, narrow flakes appears to be similar to that used in New Mexico by a culture known as the Clovis, which is thought to be some 13,000 years old. If the Clovis tools, which would have been used to kill such beasts as mammoths and bison, weren't developed independently, then the Solutreans might have brought them to the Americas first.

This version sits at odds with most other academic accounts of how the Americas came to be populated by our species. Geologists know that less than 15,000 years ago sea levels were low, allowing for a land bridge across the Bering Strait that would have joined modern-day Russia and Alaska. People could even have lived in the region between Asia and Alaska for an extended period of time before spreading further eastward. According to more recent research, there may have been waves of migration in both directions, with some people returning to Asia. So the most convincing account of what happened is this, that the first Americans came from the west, not the east, from Asia, not Europe.

The Solutrean hypothesis claims the opposite, that Euro-peans occupied the Americas long before the colonists of the seventeenth century, that in fact they were among the first people to live here. It suggests that those who became the native Americans must have reached there from Europe perhaps via Greenland and Canada, somehow crossing the treacherous

Atlantic during the last ice age, which ended around 11,700 years ago. Way back then, vast swaths of the planet would have been covered in sheets of ice, and sailing – or perhaps snowboarding – 6,000 kilometres across the Atlantic would have been a survival challenge of epic proportions. Yet those who defend this account believe that it was possible, especially if there was a continuous ice shelf across the ocean, which could have provided fresh water and food throughout the journey.

It's a theory at the very margins of science, but there remain a small number of American archaeologists who have staked their careers on it, publishing books on the hypothesis and clinging to the belief that more evidence will eventually prove them right. Among the most vocal is Bruce Bradley, usually based at the University of Exeter in England. It was in the 1970s that Bradley became aware of similarities between ancient stone tools dug up in northern Spain and those found in New Mexico. He couldn't believe that these similarities were mere coincidence.

'The basic underlying technology, the way stone tools are made, unless you understand how many detailed choices you have when you're making stone tools, things seem like they could happen accidentally . . . It's not just the blades, it's the way they made all the other tools. Virtually all of them have correspondences that are very, very striking between Solutrean and Clovis,' he tells me over the phone from Colorado.

The political implications of the Solutrean hypothesis are clear. If correct, it could be read as a suggestion that Europeans had a prior claim to the Americas, because their ancestors were already there many millenia before Columbus arrived in 1492. When they came later, then, they were only returning to a land that was already theirs. 'I see it as intimately tied up with the idea of Manifest Destiny,' explains Raff. This was a belief, particularly popular in the nineteenth century, that the European settlers who colonised what became known as the United States were somehow fated to expand their dominion across North America, that it was written into their history before they even arrived. It's a narrative they thought gave them a moral claim to the land, and later helped to square the inhuman treatment and murder of Native Americans with the squeaky-clean founding values of the United States.

That said, evidence for the Solutrean hypothesis is thin, and getting thinner all the time. One of the glaring snags is that the two cultures, the Clovis and the Solutreans, existed so many thousands of years apart in time. Nobody has discovered any ancient bones in Europe belonging to people who would have followed the Solutrean culture, only archaeological traces, such as the objects and art they left behind. So it's impossible as yet to connect modern-day Native Americans to Solutreans through their genomes.

Recent genetic evidence does show that almost all modern-day Native Americans have a shared lineage, and that this can be reliably traced to people who once lived in eastern Siberia. The 12,700-year-old remains of a Clovis boy found in Montana have shown him to be more closely related to all indigenous American populations than any other group. Raff explains that since ancient eastern Siberians were also related to the ancestors of modern-day east Asians, the obvious picture of migration is that the very earliest people to land in the Americas must have travelled through Asia and come by crossing the Pacific, not the Atlantic. Then again, archaeology is a field in which it's difficult to ever be completely certain of anything. New evidence can emerge at any time, overturning everything people thought they knew about the past. Science more broadly almost always leaves room for doubt as well. Proving something definitively wrong is tough because it requires you to look at every possibility in the universe, and then rule them all out. Sometimes this can be done – the earth is round and it rotates around the sun, we know that for sure. But when it comes to studies of the past, it's notoriously difficult. There's always the chance that a skull will turn up from under a ploughed field, or that a fresh scrap of archaeological evidence will bubble up from the Atlantic. Anything can happen. So, hypothetically, anything could be true.

This space for uncertainty, sometimes so small that you need a microscope to see it, is where the controversies live. And as far as Bruce Bradley is concerned, however controversial the Solutrean hypothesis, there's also a chance that he might be right, that history will dig up the evidence to vindicate him. 'Disproving is very, very difficult, and I don't even like the term prove and disprove. It's a matter of probabilities. Is this evidence more likely to indicate this than that? And that's the way we work all the time.'

Since he started working on the hypothesis a couple of decades ago, Bradley has come under sustained criticism from fellow archaeologists and geneticists. One team of researchers in the United States has even described his position as 'Solutreanism', implying that Bradley and those who share his ideas have crossed the line from science into ideology. Jennifer Raff insists that the lack of evidence linking Solutrean culture to Native Americans is about as clear as it could be. 'You would expect to see a bunch of other technologies,' she says. 'You would expect to see cave art of the same kind, you would expect to see settlements, and you don't see any of that.'

Geneticist Mark Thomas agrees that the theory has only the slightest likelihood, if any, of being correct. 'Let's be clear,' he says, when I raise the subject with him, sounding surprised that I'm even mentioning it. 'This is not a scientifically prevalent idea at all. If you are going to measure weight of argument in terms of word count, then maybe it seems prevalent. But this is very much like saying climate change is controversial because there are lots of words written saying there's no such thing as human-driven climate change. No.' For him, this is about a handful of researchers who have become so attached to an idea that they have embarked on a 'confirmation bias odyssey', as he calls it, scouring the world for evidence while neglecting whatever doesn't fit.

Despite my best efforts to put him at ease as we speak, Bradley does get combative, occasionally even raising his voice during our interview. 'I'm not trying to make anybody believe this hypothesis,' he tells me. 'I'm just putting the evidence out there and saying what we think it means.'

*

In a 2010 self-published novel, titled *White Apocalypse*, white-skinned Solutreans, having crossed the Atlantic and settled in North America, are slaughtered by savages who later cross the Bering Strait and become today's non-white Native Americans. The author, Kyle Bristow, a Detroit lawyer active on the political far right in the United States, makes fictional heroes of real-life archaeologists like Bruce Bradley, painting them as victims of a conspiracy by Native Americans and liberals, who don't want to face the apparent truth that the original Americans were white. Bristow's book has become popular in white supremacist circles. One review stated, 'This evidence could be the jolt whites need to awaken from our suicidal slumber.' When it was republished in 2013, he even included what were described as supplementary materials showing the validity of the Solutrean hypothesis.

For scholars such as Jennifer Raff, this comes as small surprise. The Solutrean hypothesis speaks to a nineteenth-century worldview that painted Europeans as the true inheritors of America, the only ones capable of civilising the continent. At the time, evidence of sophisticated technological cultures such as the Maya, Inca and Aztecs, which existed long before the Europeans arrived, were only further fuel for confusion. 'Since the beginning of the Americas, there has been this question of: Who are the Native Americans? Who are they? People actually wondered, are they *humans*? The first colonists did not really have a way to incorporate them into their biblical worldview,' Raff explains. 'After their humanity was more or less accepted, it then became this idea that, well, are they responsible for creating the culture, the very sophisticated technologies and art and monumental architecture that we see?'

In the shock of uncovering complex ancient civilisations in the New World, the first Europeans imagined elaborate ways in which they could have got there. 'It's so interesting to me when I look at ideas, alternative ideas, to explain the archaeology,' Raff continues. The Solutrean hypothesis, she says, is just the latest iteration. 'People are so desperate to find a non-mainstream answer to a lot of these issues. They won't just invoke Europeans, they will invoke aliens! They'll invoke people from Atlantis! Whatever they can find, as long as it's not Native Americans.' When *The Book of Mormon* was published in 1830, it claimed that Native Americans were descendants of the lost tribes of Israel who migrated to the Americas around 600 BCE and had been cursed with a dark skin for slaughtering their righteous relatives.

Of course, just because theories are exploited by the far right, this doesn't necessarily make them false. Raff is quick to add that while some supporters of the hypothesis may be motivated by racism, she doesn't believe that the researchers themselves are driven by this. 'They are good scientists and they are legitimate scientists. They are very well respected.' At the same time, though, she sees a doggedness in them that sets them apart. 'Everybody I've talked to who actually knows them personally tells me that you cannot change their minds. Nothing will change their minds. Nothing,' she adds. 'I wouldn't go so far as to call the Solutrean hypothesis pseudoscientific exactly. It did start out as a legitimate area of investigation, but I see it right now as being almost more ideological. I mean, people are not accepting any evidence against it. If you're pro-Solutrean, that's it.'

Of course, Bruce Bradley sees it differently. He tells me that Raff is 'deluded, to put it bluntly'. He believes he's been marginalised by the mainstream scientific consensus not because he is blindly clinging to a discredited idea, but because he is brave enough to challenge the academic orthodoxy. As far as he's concerned, his detractors are the ones motivated by bias. 'For me, it comes down to a lot of political stuff. When I first started promoting – not promoting, suggesting – this as a hypothesis, I was working in France and Spain, and different places over there. And I had very, very strong negative reactions from different people among colleagues in Spain. I think it's colonial guilt.'

Bradley insists that his work is just good archaeology. 'We've made it very clear all along that we're talking about our thing as a *hypothesis*,' he tells me, sounding more than a little worn down. 'People need to look up the definition of a hypothesis. A hypothesis cannot be right or wrong.'

The debate intersects with another recent controversy, that relating to one of the few ancient skeletons found in North America, known as Kennewick Man. Dated at around 8,500 years old, his middle-aged bones were discovered in 1996 by college students in Kennewick in the state of Washington.

Researchers at the time were quick to spot that his skull didn't look particularly like that of other modern-day Native Americans. In fact, one archaeologist described it as looking 'Caucasoid'. Of course, one way someone with Caucasoid features could have ended up here all that time ago was if some form of the Solutrean hypothesis was indeed correct, that the ancestors of Kennewick Man had travelled across the Atlantic from what is now Europe. A reconstruction of his face – cast in off-white, although nobody knew his real skin colour – even weirdly resembled the English actor, Patrick Stewart, best known for playing the captain of the starship *Enterprise* in *Star Trek: The Next Generation*.

Meanwhile, local Native American tribes rushed to claim him more plausibly as their own, insisting that he must have been a direct ancestor or related to their ancestors. In their historical legends, the land on which they lived had been their home since the beginning of time. They were products of it, not migrants to it. This narrative insisted that Kennewick Man simply couldn't belong to anyone else, and so, having been dug up and manhandled, he deserved a proper burial, conducted by the tribes. At the time, this call to return such remains to indigenous communities seemed irrational and emotional to some in the research community. A bitter court battle began, pitting scientists against Native Americans. It was about one origin story versus another.

The struggle over the remains of Kennewick Man wasn't just about identity or ritual. It was also wrapped up in a dark and brutal history that included scientific exploitation. In the nineteenth century Native American graves were often looted by anthropologists and hungry collectors, keen to claim their piece of this ancient culture before it disappeared, but with no respect for its traditions. Bones were rarely returned. The insults were not limited to artefacts and remains. As recently as 1990, blood samples from members of the Havasupai tribe who have lived in the Grand Canyon for centuries were taken by Arizona State University in the understanding that they would be used to study their risk of diabetes. In the end, without the permission of the Havasupai, the samples were also used to study other medical and mental disorders, including schizophrenia. The university agreed to pay $700,000 in compensation. When Native Americans defended the bones of their ancestors, then, they weren't only laying claim to their culture, they were also standing up for their rights over their own bodies.

Even so, in 2002 a judge finally ruled that the Kennewick bones weren't necessarily related to those of any modern tribes, in large part because researchers had declared that Kennewick Man didn't really look like the average Native American. As Kim TallBear in the Faculty of Native Studies at the University of Alberta has written, they privileged 'genome knowledge claims over indigenous knowledge claims'. With this, scientists were given a green light to study the skeleton.

Then slowly came the revelations.

Researchers in Denmark, led by Eske Willerslev, a pioneer in population genetics and ancient DNA, revealed in 2015 that Kennewick Man was after all more closely related to contemporary Native Americans than any other group. The tightest genetic link was found to be to the local Confederated Tribes of the Colville Reservation, which had originally claimed him as an ancestor. The indigenous groups had been right all along. He was one of their own, as much as it is possible to be when you are separated by millennia.

In February 2017, under legislation signed by President Barack Obama, Kennewick Man, now known by tribes as the 'Ancient One', was finally laid to rest in a traditional burial near the Columbia River, which runs through Washington state. The act of rewriting the story with something closer to the truth, and then returning his remains to the tribes, carried layers of significance. 'A wrong had finally been righted,' a spokesperson for one of the confederated tribes told the *Seattle Times* when the Ancient One was buried. A fresh forensic reconstruction showed a face starkly different from the first. Like Briton's Cheddar Man, whose facial reconstruction went from white to black in the space of a century, Kennewick Man, too, was completely different the second time around. Now he was given long hair and dark skin. The resemblance to Patrick Stewart was gone.

It was a lesson in how much culture and politics can shape how people read scientific evidence. It's an easy mistake to project contemporary racial parameters onto the past, explains Deborah Bolnick, an anthropological geneticist based at the University of Connecticut. She sees it happen in her own field all the time. 'If you see the genetic markers today that are found in western Europe, people will see those in the past and continue referring to them as western European, even if they're then also found in Siberia.' It's another example of an 'indexing problem', when the first available body of evidence influences subsequent thinking. Western researchers tend to have more access to European data because it's on their doorstep, so later discoveries elsewhere in the world are often interpreted relative to these.

Bolnick tells me of the example of a skeleton of a four-year-old boy discovered in south-central Siberia and thought to have been buried there some 24,000 years ago. In 2013 this became the oldest modern human genome yet sequenced, and scientists learned that he shared some genetic variants with people in western Europe. 'The way this got framed was: you have this individual in Siberia who has these western European genetic markers, and so maybe this means that there was a migration from western Europe to Siberia,' she tells me. In reality, the more parsimonious explanation, especially given the age of the skeleton, was that it was an east-to-west movement, not the other way. In other words, people in western Europe had *Siberian* genetic markers.

'Underlying assumptions and ideas definitely get embedded in ways that we don't even think about consciously, which can play out in the science,' Bolnick adds. 'We interpret data. We bring our perspectives, our framings to the data. You can use the same data to say many different things. I think modern genomic data provides the perfect example of that, because you can have different people who are all very smart and understand the data, who look at the same datasets and describe them in polar opposite directions.' It's impossible to escape our beliefs, our upbringing, our environment, even the pressure of wanting to be correct, when it comes to interpreting the facts. Our stories get in the way.

Indian historian Romila Thapar writes, 'In contemporary times we not only reconstruct the past but we also use it to give legitimacy to the way in which we order our own society.' Jennifer Raff believes this is quite clearly at play when it comes to the Solutrean hypothesis, just as it may have been when it came to understanding the true ancestry of Kennewick Man. There are powerful reasons why researchers may want to believe their own story is right, even when evidence declares otherwise.

The past is always at the mercy of the present.

*

'I remember during the Yugoslav Wars, I was in Paris.'

Kristian Kristiansen, a senior professor of archaeology at the University of Gothenburg in Sweden, works with ancient DNA expert Eske Willerslev, who carried out the investigation into Kennewick Man. Kristiansen is infectiously enthusiastic about this powerful new field of science. But as a longtime archaeologist, he also has a sober and measured perspective on the past. He agrees that leaps in genetics have the power to overturn everything we thought we knew about ourselves. They certainly challenge racial stereotypes by showing us just how much we have always mixed together throughout the past, and how much we have in common. But at the same time, he warns from his own personal experience that the political power of such insights has its limits.

It was in Paris in the nineties that he saw this for himself, during violent ethnic conflicts between Serbs and Croats in the struggle for independence in former Yugoslavia. 'I was a visiting researcher and I was living together with some expelled, you could say, archaeologists from Yugoslavia,' he says. The lives of ordinary people meant little in that place at that time. In the push for territory, a programme of ethnic cleansing, mainly of Bosnian Muslims, led to hundreds of thousands of civilians being forcibly displaced, women systematically raped, and murders so numerous and methodical that they rose to the level of genocide. Political leaders deliberately rewrote history to cast some ethnic groups as having a claim to certain tracts of land.

Reputable historians and archaeologists found themselves fighting an intellectual war against nationalist ideologues who wanted to justify their actions by promoting false versions of the past that suited their cause. 'And nobody wanted to listen,' says Kristiansen. 'That was the shocking thing. Nobody listened. And they published in newspapers, they did everything they could to get it across, but in the heat of the whole thing, they failed.' When push came to shove, truth became victim to politics. The facts mattered only if they suited the power-hungry agenda. 'Suddenly things can turn from left to right in a split second when politics changes,' he says, clicking his tongue.

This was nothing new. It had been seen before, most obviously earlier that same century when the Nazis pulled together a miscellany of racial theories to defend the genocide of millions of Jewish people and members of other groups during the Holocaust. Then, too, mainstream scientists and archaeologists found themselves marginalised and sacked while those whose ideas favoured the regime found themselves promoted and celebrated.

Bettina Arnold, a historian and professor of anthropology at the University of Wisconsin, Milwaukee, has researched just how gross these intellectual abuses were in the years leading up to the Second World War. After their country's humiliating defeat in the First World War, many Germans were looking for ways to rehabilitate their national pride, and the search for a more glorious prehistory was one means to that end. By promising to mend this collective feeling of bruised self-respect, the Nazis managed to gather public support. Slowly, they harnessed archaeological evidence that fitted their account of a great 'Germanic' past. At the same time, by proving that the German people had roots across Europe, they could lay moral claim to territory beyond their own borders. In their minds, they would expand to form an empire based on the ancient Germanic race, which they believed itself originally stemmed from noble, light-skinned Aryans, and was physically and mentally superior to all others.

Their intellectual framework came partly from linguist Gustaf Kossinna, who had been appointed a professor at the University of Berlin in 1902 and went on to become one of the country's most influential thinkers. By the time the Nazis came to power, Kossinna was dead. But the Third Reich had already nurtured his theories, seizing upon his argument that culture and ethnicity were wrapped up in each other. His ideas implied that when archaeologists uncovered evidence of shifting cultures, they were also seeing evidence of migration. So if they could find archaeological proof that the same cultures they could see in Germany had existed elsewhere, this would be proof that ancient Germans had lived there, that this was also part of *their* rightful territory. Archaeology, folklore and anthropology combined in service of this political idea.

The Nazis, says Arnold, were bent on 'proving that there was some kind of genetic – racial, essentially – commonality'. It was about expanding the boundaries of the traditional homeland using race as a rationale. This is not to say that the idea was welcomed, or even widely accepted. Kossinna was heavily criticised within his own lifetime for the quality of his work, most notably 'for the kind of cherry picking that he engaged in', she adds. 'You pick certain parts of material culture that support your arguments, you ignore those that don't. This is obviously a danger anyway in any archaeological interpretation. In his case, it was quite easy to pick holes in the arguments that he was making, and people did. Even his contemporaries did.'

Part of the reason that Kossinna was drawn to the Nazi party as they were beginning to claw their way to power in the early twentieth century is that he found support there he didn't necessarily have from his peers, reputable historians and archaeologists. 'He was a bit of a marginal figure early on in his career . . . He had been rejected by the mainstream cultural historians of the day. He had a hard time finding an academic job. There is a lot of personal bitterness,' she explains. But in the Nazi party, 'he found a niche and a place where he could matter, where his work was accepted and seen as important.'

Kossinna wasn't working for the party when he first developed his theories, but he was certainly motivated by an ethnically charged worldview that became useful later on. By the end, the party turned him into an icon, a founding father for the regime.

The politics suited him as much as he suited the politics. All the way up to his death in 1931, Gustaf Kossinna was firmly on board with Nazi ideology. Many of his publications make clear that he was aware of the political ramifications of the research he was conducting, says Arnold: 'He fully supported the idea that archaeology should be a handmaiden of the state.' Arnold has noted that in the first two years after Adolf Hitler came to power, eight new academic chairs in German prehistory were created. History was deliberately rewritten and appropriated by the party. The infamous swastika we associate with the Third Reich was employed after German archaeologists found the same prehistoric symbol on old German pottery. The 'SS' double lightning bolt that featured on Nazi uniforms was similarly adapted from an old Germanic rune.

Everything was recast through a political prism. Archaeologists writing for mainstream journals were replaced by those who toed the party line, and Germanic cultural influence on Western civilisation was intentionally exaggerated. In one bizarre instance of wishful thinking, the ancient Greeks were painted as ethnic Germans who had long ago somehow survived a natural catastrophe before developing a sophisticated culture of their own in southern Europe. Hermann the German, the tribal chief who had led his army to victory against the Romans almost two thousand years earlier, was dragged into service, too. Under the Nazis, his statue – erected in the Teutoburg Forest in the nineteenth century – became a focal point for nationalist pride, a reminder of a golden age of heroism.

Gustaf Kossinna remains a cautionary figure for archaeologists, as he does for academics more widely. The problem
throughout, Arnold argues, is that archaeology – with its shortage of evidence and abundance of interpretation – has always
lent itself to misinterpretation. The same may be said of other
scientific fields, especially when data is thin on the ground and
there are plenty of people desperate to speculate on the meaning and significance of what little there is. This has certainly
been the problem with race science, and the study of human
variation.

Kossinna is a reminder that shaping evidence around ideology, selecting specific results to suit a narrative, or even just a
lack of care when it comes to interpreting or presenting data,
can lead to disaster. What Kossinna did was no different from
how scientific information was manipulated by anti-abolitionists
in the American South in the nineteenth century, clinging
tightly to their slaves, or by British imperialists who made the
case for colonial rule by framing themselves as racially superior. But today it's a lesson taken seriously in Germany. When
the two thousandth anniversary of the Battle of the Teutoburg
Forest rolled around in 2009, the celebrations were sober, with
a marked shortage of volunteers wanting to play the role of the
Germans in a battlefield re-enactment. Most volunteers wanted
to be Romans. Showing a distinct lack of nationalist fervour,
even a spokesperson for the local museum told a reporter, 'I
hope people in the future will take a closer look at history, question what they have learned and review the sources.'

'There is in Germany among my fellow archaeologists a
really high sensitivity towards political misuse, towards simpli-
fication,' says Kristian Kristiansen, 'because they have seen the
way that the Nazi regime constructed a false prehistory, by taking
elements of the established prehistory and then twisting them.'

It was a sensitivity that could be seen in 2015 when Harvard geneticist David Reich was working on a paper for publication, examining evidence of the very same prehistoric culture that Kossinna had once described as Germanic. A German archaeologist who had supplied the team with skeletal samples was so concerned that the same conclusions might be reached about links between migration and cultural change as the Nazis had made that he and a number of other colleagues asked for their names be taken off the list of authors.

There is good reason to be cautious. In spring 2018 the prestigious science journal *Nature* issued an unusual editorial stating that, in a situation reminiscent of Gustaf Kossinna, 'Scholars are anxious because extremists are scrutinising the results of ancient-DNA studies and trying to use them for similar misleading ends.' It was the kind of warning that would have been unthinkable in a scientific journal a decade ago, but a new political climate combined with fresh discoveries emerging from human genetics was creating a crisis. 'They worry that DNA studies of groups described as Franks or Anglo-Saxons or Vikings will reify them . . .' People out there, the editorial suggested, are actively abusing science for racist purposes. In 2018 the *New York Times* reported that white nationalists had been seen 'chugging milk' at gatherings to demonstrate a genetic adaptation shared by many Europeans that allows adults to digest milk (a trait incidentally common to many non-white populations, too, who have historically also kept dairy cattle).

Kristiansen has witnessed this kind of racially motivated cherry-picking and distortion of scientific information for himself. 'Every time we publish, it goes into the global database. And what we can see is a lot of people are sitting out there that have all kinds of blogs where they go in and reanalyse data and see if they can falsify or get other results.' He suspects some of them may be fellow academics, but others seem to be enthusiastic amateurs. From what he can tell, they are deliberately scouring the genetic and archaeological data for evidence that fits in with their pet political or racial theories. In one memorable incident, he tells me that he was drawn into email correspondence with a respectable Canadian sociologist with a professorship at a public university who had cited his research. As they emailed each other, it slowly became clear to him that this man had views sympathetic to white supremacists.

'Everything can be twisted,' Kristiansen warns me. 'Everything.'

*

In the spring of 2018 a smattering of news reports began circulating in the Indian media which could have been lifted straight from Germany in the 1930s: the Indian government had set up a committee to rewrite history.

According to the reports, this was a decision that threatened to slam headlong against established scientific and historical facts, promoting a mythical version of history that painted India's dominant faith, Hinduism, as being central to its entire past. This particular origin story had been around for a century or so, enjoying varied levels of support. But it had become increasingly popular in recent decades, especially with the election of a conservative Hindu nationalist government in 2014. Now, it seemed, religious identity politics was being ratcheted into a higher gear.

Appointed by the Prime Minister, the twelve people on this new committee included a former senior official with the Archaeological Survey of India, along with the Minister of Culture, who was apparently keen to introduce a 'Hindu first' account of history into schools. Established facts about evolution, migration and genetics would be thrown out of the window in favour of a firmly religious narrative, one insisting that ancient Hindu texts are based in fact, not myth, and that those of other faiths have no claim to India.

In her book on contemporary identities in India, *The Past as Present*, historian Romila Thapar explains that the idea of a Hindu homeland has its roots in the struggle against British colonialism and efforts to construct a new national identity once independence was won. Just as in Germany following the First World War, politically motivated accounts of Hindu superiority have offered some Indians an opportunity to reclaim their self-respect, assert some collective pride and build a new sense of national identity. But in the process, India's ancient past, which is far from fully documented, has become a tool for projecting notions of technological and cultural superiority. Some of the members of the government committee to rewrite history, like other religious nationalists, believe that India belongs only to Hindus, even going so far as to suggest that Hindu Indians have no ancestry anywhere else, not even in Africa where our species originated. One member, a Sanskrit scholar, reportedly believes that Hindu culture is millions of years old, an order of magnitude older than the human species.

For religious nationalists, their ancestry and religion both tie them deeply to their land. Some have absorbed old European and American theories of an ancient, noble, pure-blooded Aryan race, and claim that these Aryans did indeed originate in India, living in the sophisticated cities of the Indus Valley Civilisation towards north-western India thousands of years ago. As the Nazis saw themselves, they see modern-day Hindus, particularly light-skinned, higher caste Hindus living mainly in northern India, as direct descendants of the Aryans. It's a connection thought to be not only timeless, but one that makes them superior to everyone else on the planet. As Thapar writes, the ideologues believe that 'the Aryans of India were not only indigenous but were the fountainhead of world civilization, and that all the achievements of human society had their origins in India and travelled out from India.'

It's a version of history that doesn't withstand much intellectual pressure. The oldest settlements to have been excavated in India, belonging to the Indus Valley Civilisation between 3,000 and 5,000 years ago, confirm that modern-day Indians must be products of different waves of migration from other parts of the world as we all are, some more closely related to Europeans in genetic terms, others less so, and everyone a mix. Hinduism and its cultures, too, have changed through time, and according to Thapar, bear little relation to the earliest civilisations.

But these facts don't always seem to matter, I'm told by Subir Sinha, a researcher at the School of Oriental and African Studies in London, who has been tracking the rise of religious extremism over the years. 'All I can say is that people who used to be scientific and rational at one time will now take a view that this [account of Hindu origins] is possible.'

While certain facts are deliberately ignored, at the same time there is a desperate desire to find others that do fit the ideology. The parallels with Gustaf Kossinna and Nazi Germany are striking. Indian archaeologists, for example, have been tasked with digging up evidence of places, people and events described in Hinduism's ancient texts. The legends include tales of demons, flying machines, monkey-headed and elephant gods. The nationalists say these weren't just beautiful allegories, but hard historical details. One example Sinha gives me is that of the mystical river known as the Saraswati, which sits at the centre of much of the action in one of Hinduism's scriptures. 'One of the first things the government did when it came to power this time was to set up a task force to identify the Saraswati River,' he tells me.

'There is a kind of will to truth. We will make this to be the truth if we try hard enough.'

Not just history and archaeology, but biology, too, are deployed to support the myth. While wilfully ignoring science on the one hand, they also 'care a lot about what the scientists are doing,' adds Sinha. When geneticists release new findings about human ancestry that don't sit well with the religious narrative, they are seen to pose an intellectual threat. It's a problem that has already landed at the door of Kumarasamy Thangaraj, India's leading population geneticist, based at the Centre for Cellular and Molecular Biology in Hyderabad. As one of the scientists who helped prove that modern Indians are the product of repeated migrations, he is well aware of the controversies surrounding the work of researchers like him. When he carried out research on Indian population genetics in collaboration with international colleagues, they decided to deliberately describe ancient migrant populations not as being African, Iranian or Middle Eastern in origin, as they might have for accuracy's sake. Instead, they called them 'ancestral north Indian' and 'ancestral south Indian', in an effort to be politically sensitive. With this wording, they avoided upsetting those who believe that Hindus spontaneously originated in India.

'It has not come to that level where I have to argue with them. People talk to themselves. They never fight back to me or oppose my findings, but that exists,' Thangaraj tells me, diplomatically.

Even so, Romila Thapar, Subir Sinha and other academics have expressed strong concerns about what they see happening in India. 'Most of the politics of connections with land and nature and "we are the true people" tends to be of a fascist, right-wing variety,' Sinha explains. 'They believe that civilisations should be based on a true homeland of righteous people, which have the same religion and language.' Religious minorities, particularly Muslims, have been picked out for persecution in this increasingly charged political environment. In the worst case, an eight-year-old Muslim girl living in the Indian-administered part of the state of Kashmir in north India was taken to a Hindu temple in early 2018 and gang-raped over a period of days before her dead body was dumped in a forest. Two government ministers attended a rally in support of the men accused of the crime.

The barbarity of how nationalism manifests itself may make it feel as though facts are peripheral, that they don't really matter, not when lives are at stake, not when young girls can be gang-raped for their faith. But for the religious nationalists, says Sinha, 'the past matters a lot for them to be confident in making the claims of greatness they want to make, claims to greatness in the world but also claims to land, power, claims to the right to show down people of other religions.' It gives them privilege over the truth, a version of their social structure that they can then sell to others. It throws weight behind the fists, it gives people the sense that what they're doing is morally justified, because this is the order of things as they see it. This is how the world was created, and they are only bringing some of that order back in place of chaos.

The nationalists must turn to the past for reassurance. The past is their problem.

But then, arguably, so is it for all of us, in smaller and bigger ways. When we study our genetic ancestry, aren't we also looking for clues about who we are, trying to reaffirm a story we have of ourselves? Why does it matter to some people that their ancestors were Vikings or Egyptian pharaohs? Does being related to Genghis Khan or Edward III make one person living today any different from the next? When we claim ethnic or racial pride, what are we doing but trying to piggyback on the achievements of those who went before us? It's not enough to be who we are now, to be good human beings in the present. The power of nationalism is that it calls to the part of us that doesn't want to accept being ordinary. It tells people that they are descended from greatness, that they have been genetically endowed with something special, something passed down to them over the generations. It attaches them to origin stories that have existed for hundreds of years, soaking into their subconscious, obscuring truth with the dazzling light of myth and legend. They are stories that shape even the convictions of world-famous geneticists like James Watson, who despite everything he has learned through science, clings to the belief that certain groups of people are simply born superior to the rest.

9

Caste

Are some races smarter than others?

ON A SMOGGY JANUARY DAY I set out by taxi from the Indian capital, New Delhi, into the nearby state of Punjab to visit my extended family. It's a journey I've taken many times before, usually napping my way past the lush farmland and fruit sellers flanking the highway, stopping only at a *dhaba* for a butter-soaked lunch before returning to sleep. But this time I stay awake to study the faces I pass while I'm on the road, to watch the bodies jostling through the traffic.

I stare and I compare. The word 'brown' doesn't do any of us justice. Every possible skin colour is represented: ebony, paper white, yellowish, and countless other shades, along with almost every possible feature. India is unparalleled. The sheer span of the country and its environmental variety, from the sun-drenched beaches of the south to the snowy Himalayas of the north, seems to be mirrored in the physical diversity of its people.

An encounter the same morning had made me look with fresh eyes at this place I thought I already knew. I'd met geneticist Sridhar Sivasubbu in his scrupulously tidy office at the Institute of Genomics and Integrative Biology on Mathura Road in the heart of the city. Part of Sivasubbu's work is to investigate human genetic variation within India, with the aim of battling rare diseases. In a nation of more than a billion, he told me, no rare genetic illness is actually all that unusual. Rarity becomes a relative concept. Someone somewhere is comfortably beating the odds. But what fascinates him more than that is the variety. India is a microcosm, an entire hemisphere represented inside one country.

'We have something like fifty-five populations, major populations. Then there are minor populations within the country, and five linguistic groups are there,' he explained. Regions overlap genetically with other nations in south Asia and parts of the Middle East. The Andaman Islanders have close genetic affinities with Aboriginal Australians. This breadth of difference may explain why today India is the only remaining country in the world that has its own government-funded Anthropological Survey, designed to study the biological and cultural variation of its citizens.

But there's more to all this wondrous human diversity than meets the eye. One of the unsettling reasons that Indians exhibit such physical difference is that it's partly self-imposed – the culture demands it. Many centuries of marriage within fiercely tight-knit communities and a caste system that stretches back perhaps two millennia to keep privileged and non-privileged people apart, reinforced by the British under Empire, have deliberately separated populations from one another to this day.

Historian Romila Thapar has noted that there was always intermixing between population groups in India, that the strict divisions evident today were probably less strict in the past. But unlike neighbouring China, which though larger is not quite so ethnically diverse, freedom in India to marry and move between groups does seem to have been restrained. Millions still prefer to marry within their own religion, colour, caste and community, however shallow a pool of potential partners that might leave them with. And their preferences are policed not in law but by families, with regular cases of couples being attacked or killed for falling in love inappropriately. Inter-caste marriage was legalised in 1954, yet a survey in 2016 found that as many as 40 per cent of adults in Delhi who didn't belong to the lowest castes thought there should again be laws preventing it.

'You can find this in Hindus, you can find this in Christians, you can find this across India,' said Sivasubbu. 'So it's not about religion. It is about customs and marriage practices that have been passed down over generations. Indians tend to marry within the larger community that they live in, and in spite of all the few hundred years of knowledge that we have acquired, we still follow conservative and traditional marriage practices.'

My mother grew up with these values. Although she ended up married to a man of a different caste, religion and community, she retains a fatalistic view of life, steered by a society that has forever told her that everything is circumscribed, that has for hundreds of years kept people in their place. For those raised this way, social hierarchies feel knitted into their bodies. Their faith in the power of heredity is so strong that it overwhelms how they think even about themselves. They find nothing odd in a dynasty that remains in place over multiple generations, be it political, artistic or in business. Some extol the virtues of caste as giving everyone a valued place in society, ignoring that for people consigned to a life of cleaning toilets, it's little solace to be told that this is where they belong in the cosmic scheme of things. As a famous Indian economist belonging to a higher caste once told me, this is Indian culture, and it is unrealistic to expect change. It is regarded as given that people are fundamentally different, that they are born a certain way. Everyone becomes trapped in the net of their ancestral history.

Parallels have been drawn with the British class system, or race in the United States, but caste has features of both and of neither. It's ugly in its own way. At birth, you inherit your place in this social hierarchy and few transcend it. An 'untouchable' at the bottom will be given society's dirtiest work, existing in a permanent state of impurity, while those at the top of the ladder are a kind of aristocracy, favoured for jobs and education (people I meet as a journalist still like to drop into conversation that they are Brahmins, the highest, priestly caste, expecting this to carry some currency). The distant origins of all this are likely to have been partly strategic, to keep wealth and property within families. Some castes tend to encompass trades, creating generations of teachers, merchants or fishermen. Those at the peak of the system are apparently supposed to show benevolence to those below, but in reality there is considerable discrimination and violence.

Successive governments have brought in reservation policies, setting aside jobs and scholarships for disadvantaged castes. Even so, a report in 2014 by the international advocacy group Human Rights Watch found that teachers in some schools were still forcing lower-caste students to clean toilets and sit apart from everyone else. Top universities and colleges remain heavily populated by students of higher castes.

When scientist James Watson compared Brahmins to Jews, then, claiming that both had been bred for academic excellence, he was using both the language of caste and of science, seeing their differences as genetic qualities passed down over centuries. Their fortune is believed to lie beyond circumstance, to be part of what they are.

So Sivasubbu's comments overshadowed my thoughts as I rode to Punjab. In India, skin colour doesn't always faithfully betray someone's caste, but there's a lingering view that being fairer is better, that it denotes a higher status. The four main caste groups are sometimes even designated by colour – white, red, yellow and black – not unlike Europeans' classification of human races in the nineteenth century. 'Somebody very tall or very fair-coloured, obviously they would select an individual with similar features,' I had been told matter-of-factly by Indian population geneticist Kumarasamy Thangaraj. 'So there is a selection operating, not by natural but by man-made selection.' I recall when I was flat hunting in Delhi before starting my first job there as a reporter, I was asked to list my skin shade on the rental form. Having thought of myself only as 'brown' for most of my life, I had no idea what to write. The letting agent took a good look at me, and with a dirty smile scribbled, 'wheatish'. Colour takes on a new subtlety when every degree of pigmentation matters.

In both biological and social terms, India has long been a unique case study for scientists. This is why race scientist and eugenicist Reginald Ruggles Gates was so captivated by it. Here, systematic discrimination, the notion that groups of people are biologically pure and should be kept separate, that there are different breeds, isn't just an ideology. It's a living practice.

*

In the 1950s Indian geneticist L. D. Sanghvi wrote that people in his country 'are almost under an experimental environment . . . broken up into a large number of mutually exclusive groups, whose members are forbidden, by an inexorable social law, to marry outside their own group'.

When the idea of race research became unpalatable after the Second World War, Sanghvi was one of the first to turn to population genetics. In one nation, scientists like him believed they could explore what happens when human groups stay 'pure' – the kind of purity that nineteenth-century race scientists imagined might be possible, that Hitler wanted to see in Germany for the Nordic 'master race', and that white supremacists still want to see in Europe and the United States today. The grand social experiment that had already taken place in Indian society could reveal first-hand how the world would look if people mated only inside their narrow communities, selected for certain qualities over many generations.

It's remarkable just how widely Indians today believe that caste is deeply, biologically meaningful, that it has created exactly what it must have been intended to create: a social order reflected in biology, with the smartest and most gifted at the top, and others in various professions with their own skills below, as warriors and merchants, or as cleaners and servants. Even scientists think this way. 'The caste system, whatever has been practised for the last several generations, or several thousands of years, has a definite impact on everything,' Thangaraj told me. Character traits and abilities get passed down over generations, he implied. 'The offspring that's coming out of that founder are going to have such character. Usually, then, they become very unique features of that particular population.'

Sridhar Sivasubbu, too, suggested to me that people are biologically suited to the groups they are born into. By separating themselves for so long they have created genetic enclaves with particular talents, making caste not only a social reality but a biological one, too. 'Clearly certain communities have certain biological abilities which they are born with . . . we all look different and we each have our own strengths and abilities.' For him, the differences are so profound that castes are analogous to separate races, as population geneticist Luigi Luca Cavalli-Sforza might have defined them. 'You could call two groups, two peoples, completely different races and treat them as two separate entities. Or you could just celebrate both of them and say that they are different and each has the unique strengths and weaknesses.'

By way of an example, Sivasubbu pointed to Haryana, one of the states bordering New Delhi, which happens to be home to a disproportionate number of sportspeople, particularly wrestlers and runners. 'So, clearly they seem to have a better physique in terms of strength,' he suggested. Another example he offered was that of several tribal communities, which he claimed were naturally gifted at archery.

This casual speculation surprised me, coming as it did from a well-regarded geneticist. It demonstrated that more than half a century of research into human variation hasn't eliminated prejudice within science, wherever it's done. Old stereotypes are still being projected onto people, but perhaps in new ways. In Haryana there is certainly a long cultural tradition of wrestling in some families, for which people train their whole lives. But lifelong training could just as easily explain the prevalence of sportspeople as any innate ability. And if tribal communities happen to contain more skilled archers, this is most probably because they have traditionally been the ones to use bows and arrows, developing their skills through sheer hours of practice.

There's a slippery slope here, assuming that everyone in a particular community should be limited to certain paths in life. Indeed, social categories like these were harnessed and pro-moted by the British during colonial rule. My father's family, who were in the military and fought in both world wars for

Britain, were designated as one of the 'martial races'. They were seen by the British as physically and morally perfect soldiers. Yet it is a family tradition that has already disappeared. My father became an engineer, his brother a headmaster. Few of his siblings followed their ancestors into the military, and certainly none of their children.

Nevertheless, I can't help but ask myself whether Sivasubbu has a point. In 2018 scientists were amazed to discover that the nomadic Bajau people of south-east Asia, who live almost entirely at sea, surviving by free diving to hunt fish, have evolved an extraordinary ability to hold their breath underwater for long periods of time. The Bajau tend to have disproportionately larger spleens than neighbouring farmers, possibly helping them to keep up their blood oxygen levels when diving. There appears to be a measurable genetic difference between them and others, sharpened over many generations by living in an unusual environment.

This raises a question we don't like to ask out loud, but one that goes to the heart of the race debate. It is where race science began, with a belief that neglects history and jumps straight to the conclusion that the human zoo is like an animal zoo, each of us defined deep down by our stripes and spots. It follows from the offensive observation made by James Watson on the preponderance of Jewish intellectuals and Indian Brahmins in academia. Might it be possible, as Watson implied, for a group of people, isolated enough by time, space or culture, to become different? Could they, as eugenicists once suggested, evolve certain characteristics or abilities? Might they differ in their innate capacities?

Wandering down this road may be of scientific value, but it is risky, I know, and it's paved in blood. 'Could there be psychological differences between population groups?' I asked Thangaraj, tentatively. 'Differences in cognitive abilities?'

'That kind of thing is not known yet,' he replied. 'But I'm sure that everything has genetic basis.'

*

Back in London, I'm on the train on my way to a leafy corner in the south of the city, Denmark Hill.

The question of whether cognition, like skin colour or height, has a genetic basis is one of the most controversial in human biology. It's a grenade. And Robert Plomin is one of the few who have dared to handle it. A professor of behavioural genetics based at the southern campus of King's College London, he has dedicated his career to the search for the roots of intelligence, becoming one of the most divisive researchers in mainstream science. His work has far-reaching implications for how we think about human difference.

Tall, sporting a smart white beard and a crisp, pale blue shirt, Plomin is disarmingly charming in person, and rarely says anything to me that doesn't sound perfectly reasonable. I find him ruthlessly careful with his words. But there's a subtext. He moved to Britain from the United States in 1994, becoming known for brandishing the view that the cognitive differences between individuals can be accounted for largely in some way or another by genetics. The implication is that we are who we are, however we're raised. At the sharpest end, a few go so far as to take it to mean that the achievement gaps we see between large population groups (or races, as some might call them) may also just lie in their DNA.

It was in the 1970s, working at the University of Texas at Austin, that Plomin decided to dip his toe into the controversial field of behavioural genetics, eventually asking himself whether individual differences in intelligence might be heritable, and to what degree. It was perilous scientific territory with enormous social implications.

For him as a psychologist, 'it was still kind of forbidden to study genetics,' he tells me. 'It was dangerous professionally.' One obvious reason was the dark history of eugenics in the United States and Germany, as a result of which people were sterilised or killed in the belief that they would pass on their 'feeblemindedness' to their children. Another reason was that the handful who doggedly stuck with intelligence research after the war tended to say inflammatory things. In 1969 American educational psychologist Arthur Jensen, for example, claimed that black Americans had substantially lower IQs than white Americans, and that IQ was also significantly heritable. This implied that the black–white intelligence gaps seen in some tests weren't because blacks were socioeconomically worse off or discriminated against but were due to some innate genetic weakness. Plomin tells me he both personally knew and defended Jensen against his critics before he died. At the time, there was no genetic evidence to support Jensen, but he predicted that scientists would one day discover 'intelligence genes', which he believed would be 'found in populations in different proportions, somewhat like the distribution of blood types'. To compare 'intelligence genes' to blood types was telling because it was already known that blood types do vary in frequency between population groups. This would lead neatly to his next bit of speculation: that these 'intelligence genes' would also be found in lower frequencies in the black population than in the white.

Jensen's work went on to feature prominently in *The Bell Curve*, the controversial 1994 book by Richard Herrnstein and Charles Murray. Herrnstein, a psychologist like Jensen, had long insisted that intelligence was heavily heritable, he warned in 1971 that America was already slipping into a genetic caste system based on intelligence. Successful people, he claimed, were marrying each other and creating more successful children, while the unemployed languished because of their intellectual disadvantages, which 'may run in the genes of a family just as certainly as bad teeth do now'. Parallels with India's historic caste system were clear, except here it was being framed in purely biological terms. In Herrnstein's vision of the world, class, wealth and race overlapped because of biology, not because of history. The rich were rich because they were smarter, and their children were rich because they, too, were smarter. If certain racial or ethnic groups were poorer, then, it was their own fault.

In 2006 Arthur Jensen and Canadian psychologist John Philippe Rushton, who had been a pivotal figure in the Pioneer Fund and the *Mankind Quarterly*, published a brief commentary in the journal *Psychological Science*, repeating what they and Herrnstein had insisted their whole lives: that there was a racial gap of around fifteen IQ points and that it was unassailable.

But as they knew, even if this gap existed in IQ tests, proving that it was genetic and not the result of schooling, nutrition, upbringing or discrimination meant separating the effects of nature and nurture. This is the conundrum that has haunted human biology for more than a century. How do we know that differences between people are innate and not simply the product of social and cultural factors? Every human being is a unique product of biology and environment, which makes answering this question almost impossible.

Since there's no way to raise human genetic clones in a laboratory that we can run experiments on, the backbone of the psychological efforts to date have therefore been twin studies. By observing the similarities between identical twins, researchers have for decades thought they might be able to discern whether certain traits were potentially more heritable than others. But twin studies, too, are tainted by a toxic past. Josef Mengele, the notorious Nazi doctor who trawled concentration camps for involuntary subjects, had picked out young twins to deliberately amputate, mutilate and dissect. He carried out the kind of research that, for ethical reasons, is usually done on flies. And scientists learned nothing from it except just how far into hell a person would go to get his results.

For a while after the war, most researchers were wise to leave twins well enough alone. In 1979 Thomas Bouchard, a psychologist at the University of Minnesota, reignited the flame. Studying 100 pairs of twins who had been separated in infancy and then raised apart, he estimated that genetic factors accounted for approximately 70 per cent of their variance in IQ, implying that the bulk of the differences in intelligence we see between people who are healthy and well cared for are decided at birth. The rest lies in other factors in the environment, such as upbringing and schooling. Bouchard and his colleagues speculated that, at least in middle-class families living in industrialised societies, 'although parents may be able to affect their children's rate of cognitive skill acquisition, they may have relatively little influence on the ultimate level attained.' The implications of Bouchard's work, which had been financed initially by the Pioneer Fund, were obvious. If some people weren't doing so well at school – African American children, for example – it was nobody's fault but their own genes.

At the time, Bouchard was vilified, picketed, called a racist. Many decades have passed since then, but in some ways Robert Plomin has inherited this tarnished mantle. He has become part of the push to rehabilitate intelligence research, running his own studies using twin and sibling data to understand inheritance. The key is to see, as Bouchard did, whether twins raised in separate environments end up the same. And he believes that, more or less, they do.

＊

'Everything is heritable,' Robert Plomin tells me straight out.

'In fact, I am not aware of anything reliably measured that's been shown not to be heritable in terms of psychology . . . Everything is moderately heritable.' Using studies of twins – especially those adopted into different families – his estimate for the portion of intelligence that is heritable sits at around 50 per cent. This may be far lower than the figure Bouchard came up with, but it is still pretty high. If half of our intelligence can be decided by our genes, then a large part of academic achievement may well be innate, immutable.

There are important caveats, however. First, measurement of intelligence is itself fraught, nobody fully satisfied that any IQ test can really do the job or, for that matter, that researchers have pinned down what intelligence really is and whether it can even be captured by a test. Theories about intelligence tend to be culturally loaded. Second, rates of heritability aren't the same for everyone. They depend critically on the environments of the people you're studying. Take a packet of seeds and shake half of them into a container filled with nutrient-rich soil, blessed with all the water and sunshine they need. Take the other half and put them in a container of poor soil with little water and light. In both pots, individual plants will grow to various heights, some taller, some smaller. The differences you see *within* each pot are largely hereditary because their conditions are the same. But in the first pot, each seed has been given the full opportunity to achieve its potential. In the second, they haven't been, so the plants will inevitably look smaller and scrappier. In this second pot, even the naturally strongest seed may not reach the same height as many of the plants in the more fortunate container. So the differences *between* the pots are not attributable to heredity.

Some traits, such as hair colour, are very strongly determined by our genes. Hair colour doesn't change depending on the environment, unless perhaps that environment is a hair salon. Even skin colour is to some extent affected by how we live. A group of paler-skinned children who play out in the sun all day will temporarily end up with darker pigmentation than paler children who don't, but this difference is purely due to environmental factors. So when scientists say that a trait such as height or intelligence is partly heritable, the only way they can know to what degree is by looking at people in the same normal, healthy environments, with few differences in how they are raised or treated by society. If people are deprived, it can obscure the genetic influence.

For instance, studies have shown that although their populations were until recently the same, North Koreans are today on average a little shorter than South Koreans. An alien landing on our planet with no knowledge of their histories might call it a racial difference, but it is purely down to their dramatically different economic circumstances. Height has very high heritability, but South Koreans have been prosperous and well fed and North Koreans haven't, so the differences in height between them are not genetic at all.

The problem with studies of adopted twins, as critics have noted, is that they usually involve children of fairly comfortable socioeconomic means. Even if the siblings are raised apart, they're unlikely to be at the poorest end of society, where the lack of good nutrition and a stable home life may be factors in their upbringing. They tend to go to fairly good schools. There's a risk, then, that these studies underestimate the role of environment across the true range of how people live.

But let's just assume for now that Robert Plomin and his twin studies are reliable, and intelligence is both measurable and highly heritable among children raised under normal conditions. Scientists like him must also be able to point to the genes responsible for the effects they claim to see. They need to be able to explain, step by step, how they get from this twin correlation to the genes and then to the brain. And to date, they haven't found anywhere near all the genes involved, let alone mechanisms for how they impact intelligence. In 2017 Plomin, along with a battalion of researchers in the Netherlands, Sweden and the United States, published the results of a study of nearly 80,000 people, which claimed to have found forty new genes linked to intelligence, bringing the total known to have such an effect to fifty-two. It was announced in the press as a breakthrough, but in reality these genes represented only a drop in the ocean. There are many, many more. 'We're not talking about a handful of genes, we're talking about thousands of genes of very small effect,' he admits. Half a century and millions of research dollars on, Arthur Jensen's prediction remains stubbornly unfulfilled.

Despite being unable to definitively isolate intelligence in human DNA, Plomin is proud of what he's achieved so far. He believes he is getting closer to an answer. 'Do you know, two years ago we could explain one per cent of the variance in intelligence with DNA? Now we can explain ten per cent! And it's only getting bigger. I would be amazed if at the end of the year we're not explaining fifteen per cent.' Supposing he does, though, he would still be left with the challenge of finding a single mechanism, one biological pathway, to explain how any of these genetic variations acts on the brain and leads to what we see as someone's general intelligence. We know, for instance, that X-linked mental retardation is a genetic condition, identifiable in a person's DNA, reliably leading to certain intellectual disabilities. There's a quantifiable link between the gene, inheritance and cognition. But for everyday intelligence, scientists don't have anything like this.

Psychologist Eric Turkheimer at the University of Virginia believes they will never find it. 'I've been around for a while,' he tells me. 'I've been in this field thirty years, and every single one of those thirty years, the biology people of one stripe or another have been saying, "I know we're not there yet but in five years, as soon as this next piece of technology is nailed, as soon as we have brain scans, as soon as the genome project is completed . . ." It's always right around the corner. And the reason I don't believe it is because I don't believe that's the way genetic causation works.'

Turkheimer compares intelligence to marriage. Psychologists know that if you have an identical twin who has been divorced, you are more likely to be divorced yourself. There is no suggestion that there's such a thing as a gene for divorce, because people understand this to be a complex outcome, influenced by countless factors, social as well as to do with personality and temperament. 'I think there are limits to how much we can understand something as complicated as divorce looking from the bottom up.' When it comes to intelligence, like most other complex traits, heritability depends crucially on context.

He and his colleagues have seen, for instance, that in studies of people with the lowest socioeconomic status, environment explains almost all the variation researchers see in IQ, with genes accounting for practically *nothing*. Children who are the most socially and economically disadvantaged have been shown to lose IQ points over their summer holidays, while the most advantaged ones gain knowledge and skills.

So for Turkheimer, it beggars belief that anyone should assume that the cognitive gaps psychologists now claim to see between groups in the United States must be biological. The effects of slavery and centuries of racism, in all its forms, are hard to quantify, but African Americans have undoubtedly suffered in ways that have left their mark on generations. 'Millions of people were kidnapped and thrown in the bottom of boats, and taken across the ocean, and a third of them died on the trip, and then thrown on plantations and enslaved for hundreds of years. And after that, treated with total discrimination. And now, *now* their IQs are a little lower? And we're saying it's in their *genes*?

'My feeling about that is give me a break.'

We know from the plant pot example that you can't contrast populations that live in different environments because the measurement of their heritabilities will be different. And there is no doubt that the social and economic circumstances of most black Americans remain significantly poorer than those of white Americans. The Institute for Women's Policy Research in Washington, DC, found that in 2017 the median weekly wage taken home by a full-time working white man was a third higher than that taken home by a black man. Compared with a black woman, it was almost 50 per cent higher. The wealth gap, which reflects capital accumulated over generations, is even starker. Research in 2017 showed that for lower and middle-income households, white families have four times as much wealth as black families. Across the board, from police brutality to quality of healthcare and schooling, black Americans are significantly worse off.

The logical consequence of insisting that IQ gaps between

races must be biologically determined is that nothing in human society can really be changed. In an age in which some like to believe that we have transcended the old rules of social inequality, when the playing field is supposed to be level, when women have the vote, when black Americans have civil rights and colonialism is over, they believe that biology is all that's left to explain the disparity that remains. Inequality, then, must be natural, the product of the survival of the fittest. Yet we still don't have the genetic evidence to prove any of this, says Turkheimer. All we have is the belief that the proof *will be* there somewhere in the genes. 'I don't see how we can get from where we are now to that kind of racial speculating that people like to do.'

*

Turkheimer lives in Charlottesville, Virginia. In August 2017 the city became the infamous backdrop to a Unite the Right rally, drawing together white nationalists, fascists and neo-Nazis from across the United States, brandishing swastikas and Confederate flags. Their march eventually escalated into violence, culminating in the death of peaceful anti-racism protester Heather Heyer, who was struck when a car drove into her and others who had braved the threat of violence to challenge the message of those on the far right. That day was described by many as a wake-up call to America. The dream of a post-racial society seemed more distant than ever.

'The synagogue we belong to is right next to the park where it all happened,' recalls Turkheimer. A letter from his rabbi told a disturbing story. 'It was Saturday morning, so there were services inside. They locked the doors while people marched up and down the street yelling, "Burn it down".'

In the days after this the scientific journal *Nature* felt the need to run an editorial reaffirming that science could not and should not be used to justify prejudice. It was a brief but remarkable statement, proving just how potent intellectual racism was seen to have become. 'This is not a new phenomenon,' it remarked about Charlottesville. 'But the recent worldwide rise of populist politics is again empowering disturbing opinions about gender and racial differences that seek to misuse science to reduce the status of both groups and individuals in a systematic way.'

Turkheimer explains that the problem is not in the data, which is so far either unclear or unsupportive of racism, but in the rampant speculation. If science could conclusively tell us that there was a biological difference in our DNA that made some groups smarter than others, then all bets would be off. There would be no more need for debate. 'But my point is I don't think we have that. And so what we're doing instead is speculating about our intuitions, speculating about people's dumb intuitions about Jews and blacks and whoever it is people like to speculate about.'

If bad intuition is the problem, it's a problem we all have. Intelligence is just as multifaceted as any other cognitive trait, but there's a widespread assumption that it is very heavily influenced by inherited natural ability. In reality, parents' IQ scores can only explain 15 per cent of the variance in their own children, admits Plomin. Exceptionally smart parents are likely to have children a little less smart than themselves because of a phenomenon known as regression to the mean, which works to bring everyone in a population back closer to the average. Very bright children are likelier to emerge from parents in the middle of the intelligence range, where most people live. This was precisely the statistical fact that made eugenics impossible.

As the *Nature* editorial noted, 'every individual is a potential exception.' Plomin himself is living evidence of just how possible it is for individuals to be unlike the rest of their family members. He was raised in a poor working-class family in Chicago, and neither of his parents was educated beyond school level. 'My sister is as different from me as you can be, in looks, in personality, and she was never interested in books. I would go to the library and get all these books, and she didn't go to university,' he tells me.

Today, there remains little doubt that there is at least some heritable component to what we perceive to be an individual's ability to reason and solve problems, to process complex ideas and generally figure things out. But it's the degree of flexibility in this hard-to-pin-down thing we call intelligence that still eats away at some in the scientific community, not to mention many on the outside who are interested in the politics of it all. The debate is reduced, as always, to a simple question of nature and nurture, to biology and the environment.

But it's not such a simple equation. Even identical twins can sometimes show very different abilities. In Thomas Bouchard's twin studies, he came across a pair of brothers who were raised in wildly different environments. One grew up to be an uneducated manual labourer and the other was highly educated. There was an IQ difference of twenty-four points between them. In the normal course of development, 'a ten or twelve-point difference between identical twins is not unusual,' Turkheimer tells me. 'So how much flexibility might there be in the system?'

*

In 1984 James Flynn, an intelligence researcher based at the University of Otago in New Zealand, raised a collective gasp in the scientific community when he announced that for nearly fifty years since 1932, American IQs had been rising at a rate of about three points a decade. Now known as the 'Flynn effect', this finding could have been interpreted as people getting significantly smarter with each passing generation. In fact, as Flynn recognised, people had simply become much more skilled at sitting IQ tests.

They were performing better, not because they had evolved mental capacities beyond those of their grandparents, but because the skills they had were being nurtured and sharpened now in ways they hadn't been before, by better and more education, more intellectually demanding jobs and hobbies. 'The period in question shows the radical malleability of IQ during a time of normal environmental change,' Flynn wrote at the end of his paper. Whatever link to intelligence that IQ tests measured saw a benefit from the cultural passage of time.

Comparing countries, he saw similar effects everywhere. Different versions of IQ tests are used across the world, with varying results. But between 1951 and 1975 Japan saw an IQ gain of more than twenty points. In Britain it was almost eight points between 1938 and 1979. Countries that were fully modern before the testing period began tended to show more modest gains than those that underwent significant social and economic change. Kenya and Caribbean nations made particularly big leaps. Scandinavian countries, on the other hand, showed a peak and then arguably even a little decline. Flynn proved that there must indeed be a great deal of flexibility in the system.

Mankind Quarterly editor Gerhard Meisenberg had told me that some countries are too cognitively challenged to prosper, that essentially they are poor because they are stupid. His only evidence was historical IQ test scores. Should anyone need it, the Flynn effect is some of the best proof yet that he is wrong. It shows that environment matters in IQ test results, even at a population level. In a paper published in *American Psychologist* in 2012, Flynn, Turkheimer and other experts suggested that, at this rate, the apparent 'IQ gap between developing and developed countries could close by the end of the 21st century'. Flynn has shown that the IQ performance of African Americans has risen faster than that of white Americans in the same time period. Between 1972 and 2002 they gained between seven and ten IQ points on 'non-Hispanic whites'.

Another easily overlooked fact in the American black–white IQ gap debate is that very few African Americans are quite as ancestrally African as we might think. In 1976, sociologist Robert Stuckert used United States census data to estimate that as many as a quarter of people listed as white might reasonably have some African ancestors, while as many as 80 per cent of black Americans were likely to have non-African ancestors. In 2015, geneticists at Harvard University and researchers from the ancestry testing company 23andMe investigated the heritage of more than 5,000 people who self-identified as African Americans, and found that on average almost a quarter of their ancestry appeared to be European.

If this is accurate, it should come as little surprise. The sexual exploitation of black women by their white owners was common during slavery. Founding Father Thomas Jefferson is thought to have fathered children by a slave in his household, Sally Hemings, who was herself of mixed ancestry. Of course, in the last century the United States has become more interwoven as the barriers between inter-racial relationships and marriage have been beaten down. But historically the 'one-drop rule' meant that anyone with even the smallest degree of African ancestry – one drop of blood – was classified as black. Today, anyone who looks at all 'black', however complex their heritage, is treated socially in the United States as *black*.

But if the biological portion of intelligence is rooted in a complex mixture of many thousands of genes, as biologists now agree it is, then it stands to reason that someone of mixed ancestry will have a mix of intelligence-linked genes from most if not all of their recent forebears. They wouldn't inherit genes from those with one skin shade and not another. And if that's the case, and there are indeed innate, genetic racial differences in intelligence, then logic dictates that they should show up in people of mixed ancestry. If, as Rushton and Jensen implied, black people are biologically less smart, then shouldn't African Americans with higher proportions of white European ancestry have slightly higher IQs?

As far back as 1936, a study of exactly this kind was published by American schoolteachers Paul Witty and Martin Jenkins. They picked sixty-three of the highest-performing black children in the Chicago public school system, and compared their IQs with the proportion of white ancestry they were thought to have according to their parents. Their results revealed no gap at all. Having more white ancestry didn't raise a child's IQ. Indeed the most remarkable student in the group, a girl with an IQ of 200, was reported to have no white ancestry whatsoever.

A similar piece of research was carried out in 1986, this time looking at black children who had been adopted into middle-class black and white families. Children who had one white parent and one black had around the same IQ as children with two black parents. What did make a difference to performance, though, was the family they were adopted into. The black and mixed-ancestry children adopted into white families had IQs thirteen points higher than those adopted into black families.

Psychologist Andrew Colman, originally from South Africa and now working at the University of Leicester in the UK, has interrogated the claims made by scientists who insist that innate intelligence gaps between 'races' are real. He believes that research like this strongly indicates that environmental factors could well account for the entire black–white IQ gap in the United States. Even for those who claim to show contradictory evidence, he notes, 'it is literally impossible to raise Black, White and mixed-race children in identical environments if racism itself is a significant environmental factor.' Being in the same school or even in the same family means little if society as a whole sees you as substandard.

Colman accuses researchers who cling to the idea, as one writer he cites once wrote, that 'negroes show some degree of genetic inferiority' of 'a form of self abuse'.

The United States is also a special case. What's interesting is how the debate over racial differences in IQ takes on a different flavour in other countries. In the United Kingdom, the group that achieves the lowest grades at GCSE level is white working-class boys, followed by white working-class girls. Scientists haven't jumped to claim that low intelligence is rooted in whiteness. There's no evidence that being white in the UK is a socially disadvantaging factor either, so by this logic it must be their socioeconomic status that's the problem. In the decade to 2016, some of the greatest progress in educational attainment was seen among Bangladeshi, black African and Chinese pupils. Girls have also historically tended to outperform boys, even though there is no average intelligence gap between the sexes. According to the founder of the Sutton Trust, which researches social mobility, it's clear that culture is at play here. There are social influences where class, ethnicity and gender intersect, and they all affect achievement.

This is a point that even Robert Plomin – whose work has been described in *Nature* as 'vintage genetic determinism' –

concedes. He acknowledges that studying group differences in intelligence is fraught because it is impossible to control adequately for the environmental effects, adding that he doesn't see any value in studying racial differences in intelligence. 'Based on what we know now, I don't see how you would do it. Certainly for black–white differences, people have tried for a very long time and I don't think we're any further along,' he tells me. 'We've had forty years of history of this, and it's just a lot of heat and no light.'

Heritability does matter, he insists before I leave his office. This position is, after all, his bread and butter. But he surprises me by adding that if you measure individual differences and find some people are a lot better than others, 'they all can achieve incredible levels of skill given culture.'

*

Sridhar Sivasubbu looked at me from behind his thick brush of a moustache.

I had asked him about his own community and what it meant to him. He told me that he came from a particular Tamil group in southern India known in ancient times for being warriors, and nowadays tending to join the military or police force. 'If you go back and read the history of my own community, largely people have been doing this for ages, so people just keep doing it. That's what they learned.' As a scientist, in his family he happened to be an outlier. I couldn't help but detect a hint of shame in his voice, or perhaps regret for the well-trodden path he had chosen not to take.

'People tend to follow what their ancestors did,' he mused. In a country where roots go so far back in time that truth and myth are often considered the same thing, connections to the past are not easily severed. They define how people live, forming the rigid framework of a sometimes precarious existence. 'Let's say we go back a few hundred years. You still had a set of beliefs, you still had a set of skills, trades that would be passed in the community. So you learned it, and you used that for livelihood. There will be communities that will end up growing a certain type of crop, in a certain type of region,' he explained. 'It helped them survive. You also had the bonding, friends, family, neighbours. So it helped you in times of crisis.'

But keeping close to a group has its costs. India's long history of community marriage has certainly left a mark in the genetic profiles of its inhabitants. These marks are most obvious when it comes to health. 'A genetic disease tends to remain in the community,' Sivasubbu told me. Studying particularly isolated communities has revealed stagnant pools. For the very smallest populations, their reluctance to marry outside the group can be deadly. The Parsis, descended from Persians who moved to India more than a thousand years ago, suffer such high rates of cancer that there have been fears they may disappear altogether.

He has helped identify other Indian families with rare genetic illnesses, including one in which siblings were covered in dried skin resembling scales on a snake, and another with a neurological disorder so severe that the parents wanted their children euthanised. These illnesses emerged for the simple reason that the communities concerned were too close-knit. People would end up unwittingly married to someone with a common recent ancestor, and if they happened to share the same recessive genes for a rare disease, abnormalities were more likely to pop up in their children. The same can be seen with Tay-Sachs, a genetic nerve disorder that is more common than average among Ashkenazi Jews, and in non-Jewish French Canadians living near the St Lawrence River. By isolating themselves, populations have formed tight bonds, but the smaller groups have also burdened themselves with higher genetic risks. Racial 'purity' comes at a high price.

I asked Sivasubbu whether he had married within his own community, or chosen to abandon that practice, knowing the risks. He told me that, despite fully understanding the genetic problems, his culture was so important to him that he found a wife from within his group. So long as he followed certain rules ensuring that he chose someone not too closely related to him, he believed it was better to stay tight to his community than to leave it. 'It was a very conscious decision, very conscious decision. It's going back to your roots, your beliefs,' he said, smiling.

For me, as a person of Indian ancestry raised in London, between two cultures and somewhat detached from both, this was something I had never fully understood. In that moment it struck me for the first time just how powerful culture can be. It can make us act against our better judgement, but it also anchors us in the world, in time and in place. Culture, and the safety and security it comes with, can have such a profound impact on behaviour over generations that to outsiders it may well appear to be genetic. It can seem to be woven into the fabric of a person, unshakeable, when in fact that same individual under other circumstances and raised in another place might behave completely differently.

When we see the effects of culture, we can't help but dream up biological mechanisms to explain it. The freshly unearthed travel diaries of Albert Einstein, written around 1922, have revealed that even he formed generalisations as he toured the world, despite being an anti-racist humanitarian. He described the Chinese as an 'industrious, filthy, obtuse people', adding, 'It would be a pity if these Chinese supplant all other races.'

Perhaps he imagined, like the builders of the old human zoos, that our differences run all the way from our habits into our bones, that to know one Chinese person is to know them all.

Biologists Marcus Feldman at Stanford University and Sohini Ramachandran at Brown University have suggested that the 'missing heritability' that scientists have for so long struggled to find in our DNA when it comes to intelligence and other complex traits may in fact be explained by the magic ingredient of culture. New scientific tools help us understand our genomes better, but they have only *reduced* the proportion of intelligence scientists now believe to be heritable. Feldman and Ramachandran ask the obvious: Why do scientists not look elsewhere for explanations?

In the same way that our parents pass on their genes to us, they also pass on their culture, their habits, their ways of thinking and doing things. And this can happen over generations. It is so sticky and persistent that it can seem biological to an observer. This is why measuring differences between groups, even over long periods of time, is laden with error. We are social beings, not just biological ones.

'The evolution of the Indian caste system is the perfect example of social determinism,' Indian biologist Rama Shankar Singh wrote in 2001. Studying the biological differences between castes, Singh started to understand it as a system defined not by evolved differences that made people better at different things, but as a set of barriers maintained by society for so long that it felt as if they were in the blood. Lives and choices were constrained by the invisible forces of culture, and so everything remained stratified. Everyone kept their position for fear of stepping out. We don't change easily.

But we can change. As societies do shift and inequalities finally flatten, then we start to see our assumptions overturned.

Stepping out of a rigid, unjust system can prove just how flex
ible we really are, just how far outside our genes our differences
really may lie. In April 2018, a study was published into the
performance of Indian bureaucrats hired as a result of affirmative action policies. The Indian Administrative Service, one of
the largest and most powerful bureaucracies in the world, is also
one of the toughest places to land a job. Of the 400,000 people
who apply, 7,500 are invited to sit a gruelling exam, of which as
few as a hundred or so will be offered a position. Controversially,
half of these vacancies are then reserved for marginalised castes,
whose slightly lower scores would usually disqualify them.

A common assumption has been that, even though they
help redress social inequality, these quotas must have an impact
on standards. If people are given a leg up to get in, then they
surely can't be as good? Some believe that those born into
lower castes are innately incapable of doing these high-status
jobs well, regardless of their actual socioeconomic position. But
when they investigated one particularly large sample project,
American scholars Rikhil Bhavnani and Alexander Lee found
no statistically significant difference in performance at all.
'Improvements in diversity can be obtained without efficiency
losses,' they concluded. Caste had no impact. Indeed, the
minority applicants who got through in the usual way, without
the quota, tended to perform somewhat better than average.

10

Black Pills

Why racialised medicine doesn't work

'AN ENGLISHMAN TASTES THE Sweat of an African', reads the caption of the black and white engraving.

Dated 1725, it documents everyday life in the slave trading post of Calabar, West Africa. Part of a work that later became a kind of instructional guide for European seafarers, the picture shows a man being examined before he's sold into slavery and shipped to the New World. There is an obvious inhumanity to it, the flavour of a cattle market. But it's unnerving also for a strange act that's taking place in central focus. Here, the tall, well-built African slave dressed only in a loincloth crouches on his knees so that a skinny-legged, fully clothed Englishman with a sword hanging by his side can reach his face with an out-stretched tongue to *lick his chin*.

In the following snapshot, a ship sets off with its cargo of slaves as families left behind mourn their loss, heads in their hands. The fatal transaction is complete. But there's still the memory of that odd lick. Academics have wondered just what the Englishman was doing. The accompanying description says he was confirming the slave's age and checking he wasn't sick, although it's not immediately clear how a lick would achieve this. We can only assume there was some method to it.

Centuries later, a young economist found this picture and came up with an alternative explanation. The licking, Roland Fryer at Harvard University has suggested, may have been to gauge the saltiness of the African's body, because being a little brinier might better equip him to handle the long sea voyage to the New World. It's an idea inspired by a scientific theory that claims black Americans, mostly the descendants of slaves, process salt differently from white Americans. The slave ships that transported people from places like Calabar to the New World would have seen immense loss of life along the way as a result of fluid depletion caused by dehydration, vomiting and diarrhoea, the theory goes. The proportion who naturally retained more salt would have fared better, producing a genetic bottleneck among the slaves who survived to the end.

This process of human evolution by natural selection on a rapid scale, acting only on black slaves, left them fundamentally different at the other end. Those who reached America were the saltiest.

Fast-forward to the present day. In the twenty-first century, this historical hypothesis has been commandeered to explain why black Americans today suffer persistently high blood pressure, commonly known as hypertension, at higher rates than other groups in the country. Hypertension is made worse by eating too much salt, but if black Americans naturally happen to retain more salt because of the legacy of the slave trade, then the theory goes that this might be the reason they suffer, rather than because of their diets. Their bodies, some scientists and doctors believe, just aren't the same. Others are more sceptical.

*

'Hypertension is probably the oldest chronic disease we know about,' says Richard Cooper, a 73-year-old public health researcher at Loyola University Medical School in Chicago. He has spent decades investigating blood pressure, stirred by his days as a medical student in eastern Arkansas when he saw patients dying of strokes while only in middle age. Part of the problem, he has found, is that it's a strangely nebulous disease. 'There are some people who argue that hypertension is not a disease, it's a condition, that it's a state, like anxiety,' he says. 'We don't really know initiating cause.' It isn't rooted in any one organ of the body, you can't run a scan or a biopsy for it. 'It just emerges out of the mists.' At the same time, it's a major killer. The World Health Organisation estimates that raised blood pressure accounts for nearly 13 per cent of all deaths worldwide. It is a simple thing to measure and it's a widespread problem all over the world. My mother happens to have it. In and of itself, it doesn't cause her any problems, but as the doctor reminds her, she needs to watch her sodium intake in order to bring down her blood pressure. Studies have shown that consuming just six fewer grams of salt a day could save two and a half million deaths from stroke and coronary heart disease worldwide a year. When I visit her for lunch, though, I can't help but notice how liberal she remains with the salt cellar. My mother doesn't need to remind me that it's a source of shame to serve Indian food under-seasoned. Habits don't change easily.

Despite how common it is in everyone, around the 1940s or 1950s American physicians began to notice more black patients than usual coming to them with hypertension. Today, in the United States, studies suggest that hypertension is almost twice as common in black Americans as it is in other groups. If you search the UK National Health Service website for the factors associated with high blood pressure, you'll see alongside salt, lack of exercise, too much alcohol, smoking and advanced age, one more: being of African or Caribbean descent. Hypertension is thought to be so powerfully correlated with blackness that UK clinical guidelines even recommend different drugs for black people and white people under the age of fifty-five.

Preventable heart disease and stroke resulting from high blood pressure are two to three times more likely to kill a black American than a white American. This mirrors death rates from other causes. The life expectancy of a black person born in the United States today is three and a half years lower than that of a white person. Almost every major cause of death and dis- ability – even infant mortality – hits blacks harder if they live in the United States. So in a country like this, where black people die at disproportionate rates anyway, the mysterious quan- tity known as hypertension has long been a vehicle for racial speculation.

Cooper saw this speculation for himself growing up in Little Rock, Arkansas, and when he was at medical school in the 1960s. 'There were patients who were getting a transfusion and would say, "You're not going to give me black blood are you?" You were constantly confronted with that,' he recalls. At the time, wards weren't integrated and black Americans had obvi- ously lower standards of healthcare. 'Growing up in Arkansas, race was very much apartheid.'

In some places, things haven't changed. Discrimination is a problem that extends all the way to the doctor's office even today. A hefty 432-page report published by the National Academy of Sciences in 2003 confirmed that evidence of racial and ethnic gaps in healthcare is 'remarkably consistent across a range of illnesses and healthcare services', despite surveys showing that most Americans believe blacks and whites receive the same quality of care. More recently, Roosa Tikkanen, a researcher at the Commonwealth Fund, a private healthcare foundation based in New York, found that black and minority patients in New York and Boston are disproportionately treated at poorer, less well-equipped public hospitals rather than in wealthier, high-quality teaching hospitals. She tells me that in some cases a public hospital may serve three times as many black patients as a private hospital located just a few blocks away.

Tikkanen suggests that institutional segregation and structural racism may be at play. Interviewing ambulance and emergency medical staff, she discovered that if someone was picked up from a low-income neighbourhood, in the Bronx or a certain part of Brooklyn, and that person happened to be from a minority background, ambulance personnel would by default tend to take them to a public hospital or a handful of so-called 'safety net' hospitals that have a history of welcoming poor and non-white patients. 'Even after you account for the fact that they have worse insurance, minority patients are disproportionately seen in the public system,' she adds.

Social disadvantage and inadequate medical care are obviously intimately linked to health and survival, yet in the case of black Americans, these factors remain curiously overlooked by researchers. In a paper published in the medical journal *The Lancet* in 2017, a raft of public health researchers including Mary Bassett, the New York City commissioner for health, warned that scientists were too often turning to biology to answer questions that could so clearly be better explained by social inequality. We know, for example, that 38 per cent of non-Hispanic black children in the United States live below the poverty line, compared with less than 15 per cent of children overall. Black people in poorer neighbourhoods live with worse levels of transportation, waste disposal and policing, and environmental hazards such as bus garages, sewage treatment plants and highways are more likely to be located near them. The areas where they live are also targets for cigarette and fast-food marketing.

'There is a rich social science literature conceptualising structural racism, but this research has not been adequately integrated into medical and scientific literature,' the authors of the study wrote. It's the elephant in the room. Of almost 48,000 articles they found on race and health, only 2,000 mentioned the word 'racism' even once.

It seems easier to believe that our bodies are different than to accept that our social circumstances are. The notion of black exceptionalism runs right through the history of American medicine, explains Cooper. 'There were very raw and basic racist ideas that were the norm in the medical school.' One common belief was that some diseases, such as tuberculosis and syphilis, manifested differently in blacks. Another was that even if black and white bodies were similar physiologically, black bodies had less value. In the Tuskegee experiment of 1932, the United States Public Health Service teamed up with researchers at the Tuskegee Institute in Alabama (at the time a college for black people) to track the effects of syphilis, deliberately denying patients antibiotics they knew could cure them. The men were observed until they died, their internal organs slowly ravaged. Only after ethical concerns were aired in the press did the study end, forty years later.

Hypertension is one of the conditions to have survived this era with its racial stereotypes intact. American doctors have continued to wonder whether the differences they see in rates of high blood pressure could be due to some intrinsic dissimilarity between races. They used to ask whether hypertension might even be a uniquely different illness in black people, connected somehow to skin pigmentation or testosterone levels, or to the heat and humidity of Africa.

As time passed and population studies were done, it turned out that people living in Africa, especially rural Africans, have the lowest levels of hypertension in the world. 'The people are skinny, they don't eat much salt, they're very active. There is no way that blood pressures cannot be low. It's just not possible. They don't have diabetes and they don't have hypertension,' says Cooper, who has carried out blood pressure studies on tens of thousands of people across Africa, Europe and the Americas. People in Nigeria and Ghana in West Africa, from where most black Americans can trace their ancestry, are known to have far lower blood pressure than those in other countries.

Topping the hypertension charts, exceeding levels elsewhere in the world, are in fact Finland, Germany and Russia. 'They have terrific hypertension,' adds Cooper. White North Americans and Canadians, meanwhile, tend to have lower levels than Europeans, including those in England, Spain and Italy. Hypertension, then, isn't a global problem for those with black skin, it's a local one. We know that black *Americans* have higher rates of hypertension on average than white *Americans*, and the same appears to be the case in Britain.

For doctors, the question has been: Why? The most obvious explanation would be the same reason that my mother has hypertension – diet, stress and lifestyle. 'Diet is the underlying cause of hypertension,' says Cooper. In Finland, for example, diets have traditionally been low in fruits and vegetables and high in fatty meat and salt. Food in the American South, traditionally associated with African Americans, is similarly rich in salt and fats. Cheap processed foods also have more added salt. Scientists seeking an easy explanation for differences in hypertension can find one in every kitchen.

But for some reason, the kitchen has been overlooked. In the 1980s, attempting to neatly square the circle of low hypertension in black Africans with high hypertension in black Americans, a doctor named Clarence Grim came up instead with what became known as the 'slavery hypertension hypothesis' – the theory, later championed by Harvard economist Roland Fryer, that black Americans are naturally predisposed to retain more salt because of a rapid process of natural selection on the slave ships that brought their ancestors to the New World.

Grim's was an evocative tale, giving the tragic brutality of slavery an extra poignancy. Sensitivity to salt, which had helped some through the brutal journey across the Atlantic, landed their unfortunate descendants in the twentieth century with the fatal scourge of hypertension. Western diets had damned them, and there was nothing they could do. The media loved it. Fans included Oprah Winfrey and the resident health expert on her talk show, Doctor Oz. With Fryer's apparent historical evidence – the picture of the slave being licked – supporting Grim's hypothesis, it all seemed to be tied up in a bow.

Not everyone was taken in by the story, though. Biologists raised eyebrows at the suggestion that evolutionary change could happen over such short timescales. Historian Philip Curtin, an expert on the African slave trade, argued that dehydration and salt depletion were not significant causes of death on slave ships. Neither, he noted, had there ever been a shortage of salt in West Africa to make people there particularly liable to retain salt. If anything, Curtin concluded, the historical evidence ran counter to the hypothesis rather than supporting it. Richard Cooper adds that salt sensitivity, if it is seen to be higher in black Americans, is likely to be a product of being primed over a lifetime for all the factors that give them hypertension in the first place, particularly diet. This is why other demographic groups who have higher hypertension, including men and the elderly, are also more sensitive to salt.

Even so, there was a widespread expectation among many medical researchers that harder proof would one day be found, most likely in our genes. If anything could settle the debate once and for all, it would be the glittering new science of genomics.

*

In 2009 researchers thought they had finally discovered the evidence they were looking for. A team led by scientists at the National Human Genome Research Institute in the United States took DNA samples from around a thousand people and discovered five genetic variants, or versions of the same gene, linked to blood pressure in African Americans. The effects were admittedly modest, but they seemed promising. In that moment it felt as though there really might be a host of tangible genetic differences between races that would help science get to the root of disproportionate ill health in black Americans.

The search for 'black genes' already had a precedent in the shape of sickle cell disease, a serious blood condition more prevalent in people who have ancestry in malaria-afflicted regions such as West Africa. The faulty gene indicative of sickle cell disease is known also to provide some resistance to malaria, which can otherwise be fatal. This is a watertight evolutionary explanation for why such a debilitating illness has persisted. But sickle cell disease also exists outside Africa, for example in Saudi Arabia and India, which means it can be found in people of different skin shades. Within the African continent, it isn't seen in high rates everywhere; it's not so prevalent, for example, in South Africa, where malaria is less of a problem. As the UK's National Institute for Health and Care Excellence states, the gene responsible for sickle cell is actually found in *all* ethnic groups.

In the United States these nuances were lost, mainly for demographic reasons. Many white Americans tend to be of European extraction, where sickle cell is rare, and black Americans tend to have West African roots, where it's more common, so it came to be seen as a 'black disease'. Once viewed this way, it reinforced existing assumptions about essential differences between blacks and whites. Two independent facts began to align in people's minds. First, that there may be different genes determining health according to race. Second, if black people suffer illness and death at higher rates than white people, could this then be genetic?

Hypertension, however, turned out not to be so straight-forward. In 2012 another team of researchers, Clarence Grim among them, tried to replicate the 2009 study, this time with more than twice as many people. They failed. They just couldn't see the same correlations.

As scientists struggled to find the genetic evidence they thought must be out there, one team led by a researcher at the Harvard School of Public Health decided to look at factors other than race that might correlate with high blood pressure. They discovered that level of education, which often correlates with income and social class, was a far better predictor of hypertension in an individual than the percentage of African ancestry they had. Each year of education was associated with an extra half millimetre of mercury decrease in blood pressure readings. A year later, a study in Cuba showed that being black or white there made no difference to average blood pressure or hypertension. Others pointed out that living in an urban environment was strongly associated with blood pressure rises, as was being an immigrant or adopting a westernised lifestyle. Richard Cooper suggests, too, that the effects of chronic exposure to discrimination could well account for some of the black–white differences in blood pressure in the few countries where they're seen.

What is clear is that researchers are nowhere near understanding all the impacts of the social factors relating to health. One 2018 study even found a possible relationship between racism in a geographical area and the health of newborns in that area. Researchers saw, astonishingly, a direct correlation between the proportion of Google searches for the 'n' word in an area and the prevalence of black babies born prematurely or with a low birth weight. They noticed a similar heightened birth risk among women with Arab surnames in the six months after the 9/11 attacks.

For Cooper, the popularity of Grim's fairytale, the slavery hypertension hypothesis, betrays how some Americans would rather have a fanciful biological explanation for racial difference than a social one. 'It has been very hard to find genetic factors that affect salt sensitivity. What has been found is not more common in blacks than whites,' he concludes. 'There is no evidence of any significant selection across the African diaspora.' Despite the substantial resources poured into finding a gene, researchers have still found no association or mechanism that can fully account for a higher prevalence of hypertension in black Americans.

The desperate hunt for 'black genes' reveals just how deeply even well-meaning medical researchers believe that racial differences in health must be genetic, even when a goldmine of alternative explanations exists. 'It's a very useful window into people's thinking,' says Cooper. 'Below the surface, there is this very strong prior bias to believe in a mechanism like that. Racialised thinking is such a deep part, just like gendered thinking is such a deep part, of our psychology that we can't just by

conscious effort free ourselves from it completely. It keeps pop
ping up in ways when we're unprepared or not vigilant.'

People wanted so much for the hypothesis to be true, to be
able to link the trauma of slavery to the trauma of black American deaths today, that they couldn't see past it to more mundane
explanations. Hypertension, he says, is a case of science being
retrofitted to accommodate race. The data, the theories, the
facts themselves, are rotated and warped until they fit into a
racial framework we can relate to. This is the power of race. It is
the power to twist science to its own ends.

*

The list of drugs available to treat hypertension reads like a
product catalogue from Willy Wonka's Chocolate Factory: there
are beta-blockers to slow down your heart rate, alpha-blockers
to relax blood vessels, diuretics to shift salt out of your body,
ACE-inhibitors that work on the hormones, calcium-channel
blockers, vasodilators, and more. And within each of these cate
gories there are different brands, each competing fiercely for
a piece of the lucrative blood pressure pie. They represent a
global market worth in the region of £60 billion. It's small surprise that every pharmaceutical company wants to establish
its unique selling point, to mark out its drug from the com-
petition.

So it was with NitroMed, the marketers of BiDil, a pill that
combined two different existing generic drugs: one that relaxed
blood vessels and one that helped to combat heart failure, known
to develop from hypertension. Around thirty years ago it became
clear that mixing treatments in this way might allow patients
to live longer, and BiDil was one of the earliest pills to do it.
But there was a problem. The Food and Drug Administration,
the gatekeeper for medicines that can be sold and marketed in

the United States, refused to approve it because full clinical trials hadn't been done, so the company was unable to show exactly how well it worked. Its patent was running out, leaving the company in a jam. How could they get their pill approved as quickly and cheaply as possible?

The solution they came up with was unprecedented: they carried out a clinical trial in black patients only.

Studies already suggested that black patients tended to respond a little less effectively to ACE-inhibitors, for reasons that weren't yet fully understood. It's a fact so widely accepted that hypertension pills in the United States often specify different usage guidelines for different racial groups. In the United Kingdom, clinical advice goes so far as to state that white patients under the age of fifty-five should be given ACE-inhibitors to treat hypertension, while black patients shouldn't. BiDil wasn't an ACE-inhibitor, which meant it might make a promising new first drug of choice for black patients, says Jay Cohn, the pill's developer and a cardiologist based at the University of Minnesota.

Cohn's original tests on BiDil had shown that the small number of African American patients who were included in the trial (just forty-nine people) seemed to respond better to the pill than other groups did. By 2004 the results of a new trial, this time carried out on around a thousand patients, were published in the *New England Journal of Medicine*, confirming that BiDil, when taken in addition to existing medication, reduced mortality rates by 43 per cent. Given that the drug had already been shown to be effective in 1987, this outcome couldn't have been a complete surprise. But what was certainly different this time was that every single one of the patients tested was black.

Cohn tells me that this was a practical decision. 'We did not have adequate support to do a trial in the full population,' he admits. The cost of full-scale clinical trials can easily run into many millions of dollars. 'So we determined that maybe the best way to go would be to study the most responsive population, which was a self-designated black population.' This wasn't to say that BiDil didn't work in white patients, only that they didn't have the funds to do larger trials that included everyone. 'We would have needed a larger sample size to study a general population than we could get away with with a black population.' On the basis of this one group-specific trial, in June the following year the Food and Drug Administration approved BiDil as a medical treatment to be marketed solely to African Americans. It was the world's first black pill.

As soon as the decision was made, it divided people. Health campaigners and some high-profile groups such as the Association of Black Cardiologists welcomed it as a positive move, finally recognising the historically neglected medical needs of black Americans. Others, including many doctors, saw it as little more than a cynical marketing ploy to squeeze more profit out of a drug on which the patent was about to expire. NitroMed gained thirteen years of patent protection, with which it could sell the combination drug for as much as six times more than it could charge for the individual pills separately. The pharmaceutical goldmine had been excavated a little deeper.

That said, the going wasn't easy. NitroMed struggled to sell BiDil, in part because of scepticism among doctors, but also because of its eye-watering price. Since then, marketing rights have been sold to another firm, Arbor Pharmaceuticals in Atlanta, Georgia. Go to its website and you'll see black models on nearly every page, smiling and reassuringly healthy-looking. 'Although heart failure is on the rise all across America, it hits the African American community hardest,' it states. There's no doubt that BiDil is still being marketed as a black pill.

Yet, as I'm reminded by Jonathan Kahn at the Mitchell Hamline School of Law in Minnesota, who has tracked the case from the beginning, 'Race became relevant in the creation of this drug, not for medical reasons, but for legal and commercial reasons.' What's more, he explains, BiDil set a precedent. Seeing its success with the Food and Drug Administration, pharmaceutical firms began to file patent applications for other treatments that had been shown to work better in certain racial and ethnic groups. Looking at US patent applications filed between 2001 and 2005, the years before BiDil's approval, Kahn found that 65 mentioned race or ethnicity. Between 2006 and 2016, there were 384 that did.

Although no more explicitly race-specific drugs have been approved since BiDil, Kahn tells me that he has noticed an increased use of racial categories in drug labelling, the section that informs you about usage and dosage. 'You know, Asians respond differently from Caucasians, that kind of language,' he says. The website for beta-blocker Bystolic, for example, highlights how well it works in black and Hispanic patients. In the first half of 2008, Bystolic was the most heavily advertised drug in the United States. Drug labelling might seem of little importance, but he explains that it is a 'very powerful force for how medical professionals, and by extension anyone who's on these drugs, is being taught to think about the relationships of race to biology'. When you see a drug specified for use in a certain group, it implies that this group of people is biologically different from others.

In reality those working in medical research know that 'race' is hard to define, that it is a poor proxy for how human variation really works. But when there are few easy ways to distinguish people, it can feel as good as any. The ultimate aim for many in the medical profession is not to have racialised medicine, but personalised medicine, to be able to sequence an individual's genome and then tailor therapies to suit that individual. With personalised medicine, in principle, nobody will ever need to take a drug that doesn't work on them, or that gives them a bad reaction. But sequencing everyone's individual genome is expensive and ethically fraught, and we don't yet have all the data we need to analyse the results. Given these limitations, grouping people by race is seen as an imperfect but practical approximation. Most doctors and medical researchers will admit that it's a fudge, but they use it anyway. A proxy can save money and time, after all.

But really, how useful to medicine is the fudge of grouping by race? This is a question of statistics and demographics. Given all the medical data we have, how likely is it that an individual placed in a given racial group will benefit from a drug? And how likely is it that someone left out of that group and not given the drug could have benefited from it?

*

'Typically in my field, we don't just collect data and then show the data that we collected,' I'm told by Jay Kaufman, an epidemiologist and statistician based at McGill University in Canada. Data in its most raw form is rarely useful. Almost always, data has to be packaged, interpreted in some way to make sense.

He gives me a simple example. In 1970 the death rate in the city of Miami, Florida, was 8.92 per thousand people. In Alaska, at the north-west tip of the United States, that same year it was only 2.67. This is the unvarnished truth. If you happen to live in Miami, you are more likely to die than if you live in Alaska. But then, if you happen to live in Miami, you're also more likely to be retired. In 1970 the city, with a total population around

60 per cent of the whole of Alaska, had more than 92,000 people over the age of sixty-five, while Alaska had only around 2,000. If you're between fifteen and twenty-four in either of these places, the death rates are around the same.

Now take sickle cell disease. It was once suggested that only black infants in the United States be screened for the gene associated with sickle cell, because screening all infants would impose an unnecessary cost. Jay Kaufman and Richard Cooper worked together to dissect the statistics relating to sickle cell and found that, indeed, sickle cell trait prevalence in self-identified white Americans is only 250 per 100,000 members of the population, whereas in people who self-identify as black it is between 6,500 and 7,000. On this basis, it seems sensible to screen only black infants. On the other hand, there are many more white Americans than black Americans. The odds of a black newborn having the sickle trait may be 6.7 per cent, but the odds of *any* newborn having it are of the same order of magnitude, around 1.5 per cent. This is why US states today screen newborns universally, regardless of ethnicity or race.

When it comes to understanding the effectiveness of the 'black drug' BiDil, this kind of number-crunching hasn't been possible because the 2004 clinical trials included only self-identified black Americans and no white Americans. Statistical comparison just couldn't be done. What epidemiologists such as Kaufman have instead, though, are plenty of studies carried out over the decades on racial differences in response to common hypertension treatments. On the basis of these, for example, the National Institute for Health and Care Excellence in the United Kingdom recommends treating black people with calcium-channel blockers as the drug of first choice, rather than ACE-inhibitors or other alternatives.

So Kaufman and Cooper looked through all the published papers dealing with responses to blood pressure medication. Their aim was to figure out just how many individuals actually benefit from this racial distinction. They discovered that the perceived racial differences in drug response are in fact relatively small compared with differences within racial groups – exactly as anyone would expect given everything scientists know about the genetics of human variation. So while there might be statistically significant differences at a population level, this isn't always useful when it comes to treating any one individual patient. They found, for example, that regarding ACE-inhibitors, which are given to white patients under the age of fifty-five in the United Kingdom but not to black patients, data suggests that for 100 white people given the drug, 48 of them would fail to respond as hoped. Meanwhile if 100 black people were given this drug they are usually denied, 41 of them would benefit from it.

In this case, they conclude, assigning treatment by race is about as useful as flipping a coin.

Knowing that there are small average population-level differences in disease frequency, therefore, can be misleading when it comes to everyday treatment. Indeed, relying on those averages for guidance can even be life-threatening. American paediatrician Richard Garcia once described the case of a friend who as a child repeatedly failed to receive a correct diagnosis for cystic fibrosis because it was thought to be a white disease, and she was black. Only when a passing radiologist happened to spot her chest X-ray, without knowing to whom it belonged, was her condition instantly spotted. She had to wait until she was eight years old, and her colour had to be invisible, before she could be diagnosed.

For Kaufman, there are further problems when it comes to analysing the medical data on racial differences. Among the biggest pitfalls is statistical adjustment. 'The logic of these statistical adjustments is based on ideas from trials,' he explains. In experiments for new treatments, the gold standard is the double-blind randomised clinical trial, in which patients who are roughly the same in every other respect are randomly selected to receive either the treatment or a placebo but neither they nor the researchers know who has been given what. Scientists can then be sure that the effects they see are because of the drug and not something else. In real life, though, it's not always possible or ethical to carry out randomised trials, which is why adjustments are made after the fact, artificially removing the effects of variables such as age and weight.

When they do this, says Kaufman, researchers create what are in essence imaginary worlds where these things no longer matter. They are worlds made of manipulated data, illuminating a clear signal through the noise of reality. 'Once we start adjusting and describing some imaginary world, then the question is: What is that for? Why are we making an imaginary world? Which imaginary world are we going to make? What are we trying to learn from this imaginary world that we can't learn from the real world?' When it comes to racial data, the logic of adjustment is that if all the social and environmental aspects of racial difference can be removed and the data still shows a gap between groups, it must be biological.

In one study published in the *Journal of Allergy and Clinical Immunology* in July 2017, for example, Kaufman spotted an article by a large team of American medical doctors who claimed that the airways of black people become more inflamed than those of white people when they have asthma. In the United States, it's well known that black Americans carry the heavier

burden of asthma. They're almost three times more likely to
die from asthma-related causes than non-Hispanic white Americans. Black children are four times as likely as white children
to be admitted to hospital for asthma. It's also well known that
asthma is affected by the environment, including smoking,
air pollution from busy roads and factories, and living among
cockroaches, dust mites and mould. Yet this study claimed that
there was also something intrinsic to black bodies that made
black asthma patients suffer more severely, adding that they
might even need their own therapies. Kaufman decided to pick
through their figures.

The researchers' original data on airway inflammation
showed no significant differences between white and black
patients. So they chose to control for factors including lung
function and degree of control of the disease, as well as body
mass index, age and gender, until finally they came up with
adjusted data that showed there *was* a small but significant
difference between black and white patients, that black people's
airways responded to asthma in a uniquely different way.

The lead author of the paper herself tells me that her study
can't explain higher asthma rates among black people overall,
only the severity of their condition. But as Kaufman explains,
their adjustment can't even necessarily explain this. 'In a trial
we would never adjust something that is affected by the exposure.' It's like testing a drug with a side-effect of weight gain.
You can't judge the effects of that drug by adjusting for weight,
because the drug itself causes patients to gain weight. In the
asthma study 'they're saying let's imagine a world in which
black people don't have a more severe disease, then what
would their inflammatory response be? What use is that?' In
Richard Cooper's words, the science is being retrofitted to
accommodate race.

The logic of statistical adjustment holds true only if adjustment is actually possible. And racial difference is not a simple, measurable quantity like age or weight. The effects of racial discrimination, especially in a society as historically divided as the United States, run incalculably deep and wide.

'People get trained in schools to build models and make adjustments. This is the way we do things. And then they just apply it to race as though it's the same as a pill you would take. It's completely bizarre,' warns Kaufman. 'Most practitioners of medical research with medical degrees and basic science degrees don't really have much background in statistics. Many people with perfectly good intentions end up committing a lot of statistical errors because of lack of training and something we call wish bias, which is this idea that you want to find something interesting so you keep sifting through the data and fishing around until you find something interesting. That's a practice that generates many incorrect findings.'

Kaufman tells me he cannot understand why medical researchers persist in applying statistical methods to race when it's obvious that the methods cannot work the way they want them to, that they produce imaginary worlds of little or no use. 'It's epidemic in our literature that these adjustments are nonsensical. In economics, this approach that epidemiologists and biomedical people use doesn't fly at all.' He often sees scientists picking out a handful of variables to adjust for racial differences, without explaining why those variables were chosen and not others. At other times, he sees evidence of residual confounding, where the variables are measured poorly in the first place, making the final statistics even less reliable. This is especially true when it comes to adjusting for complex quantities, such as socioeconomic status.

The way data is collected and organised is also a problem. The habit of collecting data by racial or ethnic group has the unintended consequence of driving researchers to use it, hunting for gaps and trying any means possible to explain those gaps. Since 1993 the National Institutes of Health in the United States, the largest funder of medical research in the world, has had a general policy of requiring the clinical trials it supports to include women and minorities and also to collect data by 'race', across at least six categories. The purpose of this was never to look for differences between groups, but to ensure that medical studies include a broader spread of the population.

'All this is a record-keeping function that comports with other federal categories and guidelines, which are social categories,' explains Dorothy Roberts, professor of law and sociology at the University of Pennsylvania. 'It's the same categories as are used in the census to keep track of who is recruited into scientific studies. It's not a requirement for researchers to design their studies in any particular way.' But that's not the way the data always ends up being used.

In the attempt to understand racial disparities in health, social data sometimes gets treated as though it is biological data. Census categories are transformed into genetic groups. 'The US government provides millions and millions of dollars of grant money targeted to this question. If the government is giving you money, saying we want you to answer this question, and this is all people know how to do, then they're going to answer that question whichever way they know how. So that provides some incentive,' explains Jay Kaufman. It would be the same if funding agencies suddenly began collecting data by hair colour as well as by race and gender. You would be almost certain to see studies suggesting biological differences between people with brown hair and blonde hair. Just having the data invites comparisons.

As statisticians in this area know, pooling people into groups is always imperfect, and the larger the group, the more imperfect it becomes. In Britain, for example, it's common for health researchers to lump all south Asian communities into one convenient category. According to the National Health Service website, being south Asian is associated with a higher risk of cardiovascular disease, for instance. But this ignores the cultural and socioeconomic differences between Indians, Pakistanis and Bangladeshis, even within London alone. Rates of smoking tend to be high among Bangladeshis but very low among Indians, and smoking is a high risk factor for cardiovascular disease. Vegetarianism is common among many Indians, but rare among Pakistanis, and diet is also a crucial component of cardiovascular disease.

Hypertension is another condition considered to be slightly more prevalent among south Asians living in the UK. Yet although India, Pakistan and Bangladesh were one country until 1947, those of Indian origin tend to have higher blood pressure than those of Pakistani origin. Those of Bangladeshi heritage have lower levels than Britain's white population.

Although, with data as fuzzy and meaningless as this, adjusting for race or ethnicity can be a minefield, researchers routinely do it anyway. And sometimes with perplexing results, as Kaufman has found. One study he saw, for example, modelled hypertension and blood pressure for people in every country in the world in 2015 and in 1990. And 'so powerful was this model,' he wrote, 'that the authors even specified the mean blood pressure in 1990 for countries that did not even exist at that time.'

*

I ask Jay Cohn, the eminent cardiologist who invented BiDil, whether his drug, the world's first black pill, works well in patients who aren't black.

'Oh, of course it does! I use it all the time in white patients,' he replies. 'Everyone responds.'

Cohn has known this all along, and he has always been honest about it. His intentions were never to be racist, he tells me with a laugh. And I believe him. His goal was simply to get the drug approved any way he could. Labelling it as a 'black pill' was only ever driven by a commercial imperative. And in the end, this is business. Indeed, pharmaceuticals are big business.

But patients who are prescribed different drugs or who read the labels telling them that race matters may find it difficult to parse just how complex and subtle these facts are. When it comes to business, there are also those who may want to conceal them. In January 2017, independent news organisation Pro-Publica revealed that Tom Price, a Republican congressman nominated by Donald Trump to become head of the Department of Health and Human Services, had persistently lobbied on behalf of Arbor Pharmaceuticals, which owns the marketing rights to BiDil, to remove a certain study from a government website. Carried out in 2009 by heart researchers at the University of Colorado on more than 76,000 people, the study showed that across all the racial and ethnic groups investigated, the combination of drugs that made up BiDil was not associated with significant reductions in mortality or hospitalisation. The very first study on BiDil had looked at only forty-nine African Americans, and the second at just over a thousand. This was a far larger study and therefore likely to be more significant.

The US has a federal agency, the Agency for Healthcare Research and Quality, to help patients and doctors make informed choices about medical treatments. Yet according to the news report, one of Tom Price's aides had emailed the agency 'at least half a dozen times' to have the University of Colorado study removed. It turned out that Arbor Pharmaceuticals had previously donated to Price's campaign fund. The month after the ProPublica news report came out, Price was confirmed as Secretary of Health and Human Services, although he resigned before the year was out, having been criticised for his use of expensive chartered flights.

For now, BiDil is still on the market but its patent is approaching its expiration date. In the meantime, race has become a firmly and widely accepted variable in medicine, according to law professor Jonathan Kahn. He uses three clichés to describe what has happened in the last couple of decades. 'The road to hell is paved with good intentions,' is the first. 'I don't think the use of race in these contexts was nefariously plotted,' he explains. Almost everyone in the medical research community believed they were doing the right thing, curing as many people as possible in the most efficient way. The second is 'the law of unintended consequences, resulting from that. And finally, that it's creating an accident waiting to happen.' For Kahn, well-intentioned people have reintroduced race to medical science without fully understanding either their reasons or the consequences.

'It's not that the use of BiDil in black patients is immediately going to lead to the re-substantiation of slavery or scientific racism, but it's a step down the road of re-biologising race in a way that feeds deep strains of racism,' Kahn tells me. 'What we're seeing now in the US and again on the rise in Europe is a sort of ethno-nationalism. And any sort of indirect or direct approval or imprimatur of using race as a biological category becomes, no matter how well intended, dangerous in those contexts.'

And as the evidence shows, for all the studies that point to racial differences in health, the genetic evidence so far rarely tallies with them. Hypertension is just one case in point. Enormous experiments looking at the genomes of thousands of people have turned up little. Although hundreds of gene variants linked to blood pressure have been found, collectively they explain just 1 per cent or so of the variation we see, says Jay Kaufman. 'We've had a decade of genome-wide association studies now, we've spent billions and billions of dollars, and we still are at the position that it looks like ninety-seven per cent of the mortality disparity between blacks and whites in the United States has nothing to do with genes.'

This makes sense, he adds. It would be bizarre to imagine that black Americans are somehow so uniquely disadvantaged in biological terms that they would naturally die of this and almost everything at higher rates than everyone else.

Dorothy Roberts agrees that there's no logic in expecting black Americans to be so medically unusual. 'How could it possibly be that a group called black people, which first of all is defined differently around the world – it's been defined differently even within the United States, but the current definition in the United States is anyone with *any* discernible African ancestry – how could that hugely varied group, which could include someone with mostly European ancestry, someone with mostly Asian ancestry, someone with mostly Native American ancestry, how could it possibly be that that group for an innate biological reason could have a particular health outcome? That just doesn't make sense,' she says with a laugh. 'The most plausible, to me the only possible explanation could be because of inferior social conditions.'

'Race is a story we tell ourselves,' adds Richard Cooper. If you believe racial difference is biological, you will look for biological explanations. 'Everybody has a general belief in race and then they have stories about it, either something they've seen or something they've experienced. And the two reinforce each other. History, psychology, politics, we all have our belief systems and myths and things which a hundred years from now are not going to be true.'

In her book *The Social Life of DNA*, Columbia University sociologist Alondra Nelson notes that in eighteenth-century New York, the mortality rate of black infants was twice that of white. More than half the black population died in childhood, and those who lived saw their bodies driven to breaking point by the physical stresses of slavery. American medical experts claimed at the time that black people were naturally more robust, more resistant to diseases that killed others, including gallstones, tuberculosis, pneumonia and syphilis. It was a narrative that served slavery, allowing slaveholders to subject people to harsher labour and living conditions on the assumption that this couldn't harm them.

'So you have at one time a literature saying that black people are especially robust, and at another time you have a literature saying that black people are especially predisposed to illness,' says Kaufman. 'It's a contradiction, but each one serves its own purpose.'

He suggests that part of the reason Americans cling to the idea of black exceptionalism when it comes to health may be that, in some way, the idea lets society off the hook. It places the blame for inequality at the foot of biology. If poor health today is intrinsic to black bodies and nothing to do with racism, it's no one's fault. 'It says it's not our organisation of society that's somehow unfair or unjust or discriminatory. It's not that we treat people badly. It's not that we give people worse life chances,' he says. 'It's just that these people have some genetic defect and it's just the way they are.'

An interesting historical case in point is schizophrenia, a mental disorder for which people of black Caribbean ancestry living in the United Kingdom receive proportionately more diagnoses than white people, to the point where it has even been described as a 'black disease'. In recent years there has been a feverish search for the genes thought to be responsible, now that researchers widely accept that it may be heritable to varying degrees. In 2014 an enormous study involving more than 37,000 cases finally did find a number of genetic regions that may be associated with schizophrenia. But it turned out that even the most promising of these elevated risk by just 0.25 per cent. This gene variant turned up in 27 per cent of patients, but also in around 22 per cent of healthy subjects.

If schizophrenia is genetically more prevalent in one population group, then this clearly can't be a straightforward equation. Indeed environmental risk factors, such as living in an urban environment and being an immigrant, have already been shown to be at least as important to being diagnosed as any genetic links found so far. One study published in *Schizophrenia Bulletin* in 2012 found that patients with psychosis were almost three times as likely to have been exposed to adversity as children. That's not to say the disorder doesn't have a genetic component, but it does demonstrate that it can't be quantified by looking at genes in isolation. If there are racial differences in diagnoses, it may be life experiences, perhaps even the negative experiences resulting from racial discrimination, that tip some people over the edge while rescuing others. This is without even considering that schizophrenia diagnosis itself is known to be notoriously subjective.

If, even after all this, race is a factor, contrast the characterisation of schizophrenia as a 'black disease' today with an observation by Nazi scientist Otmar von Verschuer more than half a century ago. A year before the outbreak of the Second World War, he wrote, 'Schizophrenia is strikingly more frequent among Jews. According to statistics from Polish insane asylums, among insane Jews schizophrenia is twice as common as among insane Poles.' He went on: 'Since it is a matter of a hereditary disease . . . the more frequent occurrence of the disease in Jews must be viewed as a racial characteristic.'

At that moment in time in that particular place, then, it wasn't a black disease; it was a Jewish one.

11

The Illusionists

Down the rabbit hole of biological determinism

'ONCE UPON A TIME . . .'

At the turn of the millennium, excitement about the dizzying possibilities of genetics was still rife. People wondered whether gene therapy could someday cure cancer. Researchers imagined they would find genes for everything from being tall to being gay, whether we might even build designer babies by tinkering with our DNA. And two scientists working for the National Cancer Institute in the United States wrote a fairy tale.

Their protagonist is a well-meaning geneticist who one day begins to wonder why some people use chopsticks to eat their food and others don't. So of course, the hero does what all good experimentalists do: he rounds up several hundred students from his local university and asks them how often they each use chopsticks. Then he sensibly cross-references that data with their DNA and begins his hunt for a gene that shows some link between the two. Lo and behold, he finds it!

'One of the markers, located right in the middle of a region previously linked to several behavioural traits, showed a huge correlation to chopstick use,' the tale goes. He has discovered what he decides to call the 'successful-use-of-selected-hand-instruments' gene, neatly abbreviated to SUSHI. The magic spell is cast. The experiment is successfully replicated, the scientist's paper is published, and he lives happily ever after.

This might have been the end were it not for one fatal yet obvious flaw. It takes him as long as two years to hit upon the uncomfortable realisation that his research contains a mistake. The SUSHI gene he thought he had found just happened to occur in higher frequencies in Asian populations. So it wasn't the gene that made people better at using chopsticks; it was that people who used chopsticks for cultural reasons tended to share this one gene a little more often. He had fallen headlong into the trap of assuming that a link between the use of chopsticks and the gene was causal, when in fact it wasn't. The spell was lifted and the magic was gone.

Like all good fairy tales, there was a moral to this story. Although not everyone could see it.

*

In 2005 the hype around genetics had begun to fizzle out, to be slowly replaced with a healthier scepticism. Scientists began to wonder whether our bodies might not be quite as straightforward as they had thought. And then along came a young geneticist at the University of Chicago in the United States with an extraordinary claim.

Bruce Lahn's work was a shot in the arm for those who had always hoped that genes could explain everything, for the biological determinists who believed we were anything but blank slates, that much of what we are is decided on the day we're conceived. His claim was so bold it implied that maybe even the course of history could be decided by something as tiny as one gene.

Lahn had originally emigrated from China to study at Harvard University, and soon gained a reputation as a cocky maverick who didn't follow instructions, who did things his own way. A while after arriving in the United States, he changed his name to Bruce Lahn from Lan Tian in homage to the legendary actor and martial arts expert, Bruce Lee. Science journalist Michael Balter describes in a profile how once, when invited to go on a two-day hike with his colleagues, Lahn turned up with nothing but a jar of pickled eggs. 'He was kinda the whizz-kid, he was kinda the darling,' Balter recalls.

The whizz kid moved up the academic ladder at lightning speed. In 1999 he was named in MIT *Technology Review*'s list of innovators under thirty-five. Then in 2005 he published a pair of studies in the prestigious journal *Science* drawing a connection between a couple of genes and changes in human brain size. He and colleagues stated that as recently as 5,800 years ago (a mere heartbeat in evolutionary time), one genetic variant that was linked to the brain among other things had emerged and swept through populations as a result of evolution by natural selection. Their implication was that it bestowed some kind of survival advantage on our species, making our brains bigger and smarter. At the same time, he noted that this particular variant happened to be more common among people living in Europe, the Middle East, North Africa and parts of east Asia, but was curiously rare in the rest of Africa and in South America. Lahn speculated that perhaps 'the human brain is still undergoing rapid adaptive evolution' – although not for everyone in the same way.

His work caused a sensation. What set pulses racing above all was his observation that the timing of the spread of this gene variant seemed to coincide with the rise of the world's earliest civilisation in ancient Mesopotamia, which saw the emergence of the first highly sophisticated human cultures and written language. Lahn seemed to imply that the brains of different population groups might have evolved in different directions for the past five millennia, and that the groups with this special genetic difference may in consequence have become more sophisticated than others. In brief, Europeans, Middle Easterners and Asians had benefited from a cognitive boost, while Africans had languished – perhaps were still languishing – without it.

Racists ate it up and asked for second helpings. After all, here was hard scientific evidence that seemed to corroborate what all those nineteenth-century colonialists and twentieth-century contributors to the *Mankind Quarterly* had always claimed, that some nations were intellectually inferior to others. Their failure to prosper economically was rooted not in history, but in nature. 'There will be plenty more results where these came from,' predicted right-wing commentator John Derbyshire in the American conservative magazine *National Review*. Lahn also attracted support from the late Henry Harpending, a geneticist at the University of Utah and co-author of a controversial book arguing that biology could explain why Europeans conquered the Americas, and also that European Jews had evolved to be smarter on average than everyone else.

But there were problems with Lahn's findings. Even if his gene variants did show up with different frequencies in certain populations, it didn't necessarily mean that they provided those who had them with a *cognitive* advantage. The variants were known to be linked to organs other than the brain as well, so if natural selection was taking place, maybe this was nothing to do with intelligence. Maybe the genes conferred some advantage that wasn't related to the brain. The hypothesis needed more evidence.

Soon after the papers were published, controversial Canadian psychologist John Philippe Rushton ran IQ tests on hundreds of people to see if the gene variants really did make a difference to intelligence or brain size in those who possessed them. Try as he might (and we can reasonably assume that as head of the Pioneer Fund at the time, he tried his hardest), he couldn't find any evidence that they did. They neither increased head circumference nor general mental ability.

Before long, critics piled in from across the board, undermining every one of Lahn's scientific and historical assertions. For a start, the gene variant he described as emerging 5,800 years ago could actually have appeared within a time range as wide as 500 to 14,100 years ago, so it may not have coincided with any major historical events. Respected geneticist Sarah Tishkoff at the University of Pennsylvania, who had been a co-author of his papers, distanced herself from the suggestion that it might be linked to advances in human culture, as Lahn had suggested.

There were doubts, too, that Lahn's gene variants had seen any recent selection pressure at all. Tishkoff tells me that scientists today universally recognise intelligence as a highly complex trait, not only influenced by many genes but also likely to have evolved during the far longer portion of human history, ending around 10,000 years ago, when we were all mainly hunter-gatherers. 'There have been common selection pressures for intelligence,' she explains. 'People don't survive if they're not smart and able to communicate. There's no reason to think that there would be differential selection in different populations.

That doesn't mean somebody won't find something someday. Maybe it's possible, but I don't think there's any evidence right now that supports those claims.'

In the end, Lahn had no choice but to abandon this line of research. 'It was pretty damaging, because a lot of illustrious researchers either couldn't replicate his original findings or did not come to the same conclusions,' explains Michael Balter, who interviewed Lahn, his critics and his supporters at the time. *Science*, the journal that published his papers, came under attack for including the more speculative portions of his work in the first place.

To be fair to him, Lahn was partly a victim of how science works these days. The big discoveries have been made, so researchers often have little choice but to drill down into small, specific areas within biology. To make a name, they need themselves and the world to believe that this little thing they're studying is significant. According to Martin Yuille, a molecular biologist at the University of Manchester, 'If you're going to do an experiment you have to be reductionist. You have to look for one of the factors that is associated with a phenomenon, and you're tempted inevitably to try to think of that factor as being a cause, even though you know it is actually an association. So you're kind of driven to it.

'It is all too easy to exaggerate the role of the one variant of a gene that you might identify as associated with a trait . . . But you need to be modest.'

In this case, the world had seen the chopsticks fairy tale play out for real. In hindsight, it seems obvious that just because a genetic change in the brain may be more common in certain geographical populations than in others, that's no basis for claiming that it could be responsible for the fortunes of entire regions. Gerhard Meisenberg from the *Mankind Quarterly* made the same assumption when I interviewed him – that the innate abilities of a country's people are what define its success, even if we don't yet have scientific evidence for it. It's an idea that has underpinned racist thought for centuries. It assumes groups fall into ranks based on immutable biological features. It has the scent of the multi regional hypothesis, implying that nature has taken different tracks, that some of us are more 'highly evolved' than others. By any measure, the intellectual leap that Lahn took was an irresponsible one. But then some people had thought he was cocky, that he did things his own way.

When I contact Lahn, now a professor of genetics at the University of Chicago, it has been more than a decade since his controversial papers were published. In 2009, undeterred by his failure, he wrote a piece in another top-flight journal, *Nature*, calling for the scientific community to be morally prepared for the possibility that they might find differences between populations, and to therefore embrace 'group diversity' in the way that societies already cherish cultural diversity. He argued that 'biological egalitarianism' won't be viable for much longer, implying that not all population groups are actually equal. He tells me that he's still 'open to the possibility that there may be genetic differences in intelligence between modern populations, just like there may be genetic differences in other biological traits between modern populations such as bodily measures, pigmentation, disease susceptibility and dietary adaptation.'

His hypothesis hasn't changed, even though he has no more evidence to support it than he did before. Yet Lahn sticks firmly to the line that he is guided by science, wherever this may take him. 'Before there is data, these are just possibilities,' he says.

'My nose follows the scientific method and data, not politics. I am willing to let the chips of data fall where they may, as any self-respecting scientist should.'

＊

New York-based sociologist Barbara Katz Rothman has written: 'Genetics isn't just a science; it's a way of thinking . . . In this way of thinking, the seed contains all it could be. It is pure potential.'

For psychology professor Eric Turkheimer, the assumption that propels race research today in all its various forms is this deterministic pathology. 'There are people out there who think in a serious way that they're going to link up gene effects, the things you see in brain scans, the things you see on IQ tests,' he tells me. They are looking for that elusive mechanism, that magic formula which will allow them to take the genomes of people from Europe, or Africa, or China or India, or anywhere else, and prove beyond a shadow of a doubt that one population group really is smarter than another. It's all there in our bodies just waiting to be discovered.

'It's a racist hypothesis,' he adds.

In 2015 sociologists Carson Byrd at the University of Louisville and Victor Ray at the University of Tennessee investigated the belief of white Americans in genetic determinism. Having studied responses to the General Social Survey, which is carried out every two years to provide a snapshot of public attitudes, Byrd tells me they found that 'whites see racial difference in more biologically deterministic terms for blacks.' Yet they tend to view their own behaviour as more socially determined. If a black person, for instance, happens to be less smart, the interpretation is that they were born this way, whereas a white person's smartness or lack thereof is seen more as a product of outside factors such as schooling and hard work. 'So they give people a bit more leeway if they're white,' he explains.

It was also interesting to Byrd that even though the General Social Survey found that white conservatives were a little more biologically deterministic than white liberals, people with this view on both sides of the political spectrum shared the belief that policy measures such as affirmative action are needed to improve the lot of black Americans. There's a slippery slope here, he warns. 'The slipperiness is that they believe that because it's genetic, they can't help themselves, that it's innate, that they're going to be in a worse social position because of their race.' In other words, they want society to help black people, not because they believe we're all equal underneath, but because they believe we're not.

'Before it was something in the "blood" and now it's in our genes,' Byrd tells me. What has remained the same over the centuries is the racial stereotyping of black Americans. Rather than being seen as social or structural in origin, which it is, black disadvantage is conveniently rendered in the new scientific language of the day, which today is genetics. 'A lot of people have become enamoured with the science . . . the mystique of things that could be embedded within our genes.'

Stephan Palmié, an anthropologist at the University of Chicago, has argued that even now 'much genomic research proceeds from assumptions it culls from ostensibly "scientific" constructions of the past . . . and eventually restates them in the form of tabulations of [gene] frequencies.' Nineteenth-century ideas about race that have gone out of fashion take on an almost magical quality when they're freshly rewritten in the language of modern genetics. Today there is technical jargon, charts and numbers. Suddenly the old ideas seem shinier and more plausible than they did a moment ago. Suggest to anyone that

the entire course of human history might have been decided
by a single gene and they'll probably laugh. But that's exactly
what Bruce Lahn did suggest in the pages of one of the most
important journals in the world. For a moment, it felt possible
because it was *new* science.

The belief that races have natural genetic propensities runs
deep. One modern stereotype, for instance, is that of superior
Asian cognitive ability. Race researchers, including Richard
Lynn and John Philippe Rushton, have looked at academic test
results in the United States and speculated that the smartest
people in the world must be the Chinese, Japanese and other
east Asians. When intelligence researcher James Flynn investigated the claim for work he published in 1991, he found that in
fact they had the same average IQ as white Americans. Nevertheless, Asian Americans tended to score significantly higher on
SAT college admission tests. They were also more likely to end
up in professional, managerial and technical jobs. The edge
they had was therefore a cultural one – more supportive parents
or a stronger work ethic, maybe – endowed by their upbringing.
They simply tended on average to work harder.

To anyone who has grown up as a member of an ethnic
minority anywhere, especially those of us economic migrants
who were told that we would have to work twice as hard to
achieve the same as white people, this will come as small
surprise. Among middle-class Indians living in the United Kingdom (the group my parents belong to) the weight of cultural
pressure has generally been on children to become doctors,
pharmacists, lawyers and accountants. These are professions
that tend to be well respected and well paid, with no shortage
of job opportunities and straightforward entry once you have
the right qualifications. They are reliable routes into middle-
class society. Medicine carries such immense prestige bias

among immigrants and their children that, according to the most recent data gathered by the British Medical Association, around a quarter of all British doctors are Asian or British Asian. This is not because Indians make better medics, of course, but because culture acts as a silent funnel. In the same way, women get channelled into caring professions such as nursing because this is what society expects. Culture moulds people, even sub-consciously, for certain lives and careers.

We forget that these stereotypes can change over time. Asian Americans are today considered a model minority. Yet more than a century ago, European race scientists saw Asians as bio-logically inferior, somewhere between themselves and what they referred to as the lowest races. In 1882 the United States passed the Chinese Exclusion Act to ban Chinese immigrant labourers because they were seen as undesirable citizens. Now that Japan has been highly prosperous for decades and India, China and South Korea are fast on the rise with their own wealthy elites, the stereotypes have shifted the other way. As people and nations prosper, the racial prejudices move target. Just as they always have.

*

'Think about what happened to all the old racial stereotypes,' Eric Turkheimer asks me.

'A hundred years ago, people were quite convinced that Greek people had low IQs. You know, people from southern Europe? Whatever happened to that? Did somebody do a big scientific study and check those Greek genes? No, nobody ever did that. It's just that time went on, Greek people overcame the disadvantages they faced a hundred years ago, and now they're fine and nobody thinks about it any more. And that's the way these things proceed. All we can do is wait for the world to change and what seemed like hardwired differences melt away and human flexibility just overwhelms it.'

But the waiting is hard. And as we wait, it remains all too easy for researchers to allow their assumptions about the world to muddy the lens through which they study it, and for the research they then produce to impact or reinforce racial stereotypes.

In 2011, Satoshi Kanazawa in the Department of Management at the London School of Economics, who writes widely on evolutionary psychology, speculated that black women are considered physically less attractive than women of other races. 'What accounts for the markedly lower average level of physical attractiveness among black women?' he blogged in *Psychology Today*, racking his brain. 'Black women are on average much heavier than nonblack women . . . However, this is not the reason black women are less physically attractive than nonblack women. Black women have lower average level of physical attractiveness net of BMI [body mass index]. Nor can the race difference in intelligence (and the positive association between intelligence and physical attractiveness) account for the race difference in physical attractiveness among women,' he continued, in the manner of a drunk uncle.

At a stroke, Kanazawa took it as a scientific given that black women are both less attractive, which is obviously a value judgement, and innately less intelligent, which is unproven. Presenting these two offensive statements unchallenged, he landed on the speculative conclusion that their unattractiveness, as he had now established it, might have something to do with their different 'levels of testosterone' from other women, again unproven. Kanazawa, whose published work has since looked at intelligence and homosexuality among other things,

had his online post promptly pulled down under the weight
of public and academic outrage. The London School of Economics banned him from publishing any more non-peer
reviewed articles or blog posts for a year.

But how did it get published at all? When Kanazawa invoked
race as a factor to explain why he perceived some women to be
more attractive than others, he was performing a sleight of hand.
He was diverting attention away from the underlying question
of where his assumptions came from, or why he was asking this
particular question. In so doing he shone the spotlight straight
onto his racist conclusion. As soon as we, the audience, accepted
his assumptions, they quickly transformed into a scientific question. For him, it was as legitimate as asking why apples fall down
and not up, or why the sky is blue. Diverted, the publisher of
his work failed to notice that his hypothesis was dripping with
prejudice. It had no rigour to it at all.

American sociologist Karen Fields has compared use
of the idea of race, like this, to witchcraft – it's the phenomenon that she calls 'racecraft'. Race is commonly described by
scientists, politicians and race scholars as a social construct, as
having no basis in biology. It's as biologically real as witches on
broomsticks. And yet, writes Fields, she sees the same 'circular
reasoning, prevalence of confirming rituals, barriers to discon-
firming factual evidence, self-fulfilling prophecies . . .' as folk
belief and superstition. It almost doesn't matter what anyone
says because race feels as real to us as magic feels real to those
who believe in it. It has been made real by overuse.

When Bruce Lahn, just four years after he was forced to
retreat from his flawed research on intelligence genes, asked the
scientific community to embrace 'group diversity', exactly what
was he asking them to embrace? As he admitted to me himself, we don't yet have the data to tell us what the differences

between populations are beyond the superficial, and even these superficial variations show enormous overlap. The chips haven't yet fallen. His plea is not for us to accept the science we have, but to accept in advance something we don't yet know. He is assuming that data will eventually confirm what he suspects, that there are cognitive differences between groups – and telling us to take his word for it. But how scientific is that? How close is it to being simply belief?

'I do science as if the truth mattered and your feelings about it didn't,' Satoshi Kanazawa states on his personal website, lacking ostensible remorse for his paper on black women. In 2018 he and a colleague at Westminster International University in Tashkent, Uzbekistan, published a paper in the *Journal of Biosocial Science*, produced by Cambridge University Press, asking why societies with 'higher average cognitive ability' have lower income inequality. Again, he started with the unproven assumption that populations have different cognitive abilities. Again, the editors failed to notice.

Among the very few researchers to have written on links between race, intelligence and the wealth of nations are Gerhard Meisenberg, Richard Lynn and Tatu Vanhanen, all intimately associated with the *Mankind Quarterly*. Together, they have claimed that Africans have an average IQ of about 70. When Dutch psychologist Jelte Wicherts investigated this figure, he found they could have arrived at it only by deliberately excluding the vast majority of data that actually shows African IQs to be higher. 'Lynn and Meisenberg's unsystematic methods are questionable and their results untrustworthy,' he concluded. Even so, Kanazawa cites heavily from their work in his own.

It's a problem that continues outside ivory towers and marginal journals. In 2013 a public policy researcher at a powerful conservative think tank, the Heritage Foundation in Washington, DC, was forced to resign after it was revealed that he had written a doctoral thesis while at Harvard University claiming that the average IQ of immigrants into the United States was lower than that of white Americans. Jason Richwine expressed the possibility that 'Hispanics' might never 'reach IQ parity with whites', ignoring that nobody considers 'Hispanics' a single genetic population group since they have such diverse ancestries. Most Argentinians, for instance, are of European ancestry, just like white Americans. Having created the illusion that Hispanics are a distinct biological race, Richwine made it real. He had performed a sleight of hand.

From this, he followed up with the suggestion that immigration policy should focus on attracting more intelligent people. Upon joining the think tank, he also happened to co-write a study which claimed that legalising the status of illegal immigrants, most of whom are Mexican and Central American, would result in an economic loss of trillions of dollars.

By the summer of 2018, as the result of a crackdown on illegal immigrants at the southern US border by the Trump administration, thousands of young children would be inhumanely separated from their parents by border patrol officers and held in metal cages in a warehouse in Texas. News reports described them sleeping under foil sheets, wailing in distress. When the children were finally returned, families feared the long-term psychological repercussions of their traumatic detainment. In January the same year, during a closed meeting on new immigration proposals held in the Oval Office, President Donald Trump had reportedly asked lawmakers, 'Why do we want all these people from shithole countries coming here?' He was referring to people from Haiti, El Salvador and Africa. He is believed to have added that the United States should be welcoming more immigrants from countries such as Norway.

*

The notion that there are essential differences between population groups, that genetically 'shit' people come from 'shithole countries', may be an old one. But the science of inheritance helped propel these racially charged assumptions into modern intellectual thought. It is the concept of genetic determinism that has made some succumb to the illusion that every one of us has a racial destiny.

In reality, as science has advanced, it has only become clearer that human biology doesn't work this way. 'We can't sidestep the fundamental problem that biological systems are *systems*, they are collections of organisations of matter that interact with each other and each of their environments,' explains biologist Martin Yuille.

Evelynn Hammonds, historian of science at Harvard University, agrees that society has too long had a tendency to jump to the biological, to believe that our differences must be innate because how else do we explain them? 'When Jesse Owens won the medals in the 1936 Olympics, some people argued that he was not a full-blooded Negro, that he was actually mixed, and so they measured his whole body. That was 1936. They measured his whole body and made arguments like this: the thighbone is within the normal range of what the normal Negroid thighbone should be,' she says. 'That's always the hot button question at the end of the day. These people are different from each other, they're fundamentally different from each other, in disease, in athletic capacity and fundamentally in intelligence. That's the narrative. It acts as a kind of animating force under the surface.'

We can't help it. We keep looking back to race because of its familiarity. For so long, it has been the backdrop to our lives, the running narrative. We automatically translate the information our eyes and ears receive into the language of race, forgetting where that language came from. 'I think that scientists, they are trapped by the categories they use. They will either have to jettison it or find different ways of talking about this,' says Hammonds. 'They'll have to come to terms with that it has a social meaning.' This doesn't mean that racial categories shouldn't be used in medicine or in science more generally. But it does mean that those who use them should fully understand their significance, be able to define them, and know their history. They should at least know what race means.

For research published in 2007, anthropologist Duana Fullwiley, then at the Harvard School of Public Health, spent six months watching medical researchers in laboratories in California. Their job was to find genetic differences in how people responded to drugs. It was a fairly young, diverse, international team, not at all stuffy or old-fashioned. And she noticed all the scientists were routinely using racial categories not only to select their subjects, but to confidently pick out statistical differences between these racial groups. So, as Fullwiley observed, she asked each scientist she interviewed one simple question: 'How would you define race?'

Not one of them could answer her question confidently or clearly. The interviews were punctuated by long, awkward pauses and shy, embarrassed laughs. When pushed, some admitted that the concept of race made little sense, that the hard and fixed census categories actually didn't mean very much. One said, 'you can only judge race to a certain degree of confidence.' Another hesitated before admitting, 'I need to think more about it.'

Fullwiley concluded that most of the researchers 'were unsure of the meanings of the race categories that they used, yet they continue to assert that there is a biological basis to them, which they will soon corroborate'. Race was their bread and butter, the entire premise upon which they were doing their research, but they were unable to tell her what it was. Their work instead seemed to rest upon a hope that if they just persisted, they would eventually come to find scientific meaning in these categories. What they couldn't yet define would then be defined.

They appeared to believe that with enough data, with enough human guinea pigs, with enough science, they could take race – this imaginary, arbitrary set of categories invented by the powerful to control the weak – and somehow make it real.

Afterword

BARRY MEHLER'S WORDS ring in my ears.

'I have a lot of relatives who survived the Holocaust,' the historian told me. 'They are prepared for things to cease to be normal very quickly because that was their experience.'

I never imagined I might live through times that could make me feel this way, that could leave me dangling on a precipice afraid for my future. Politics is moving at such breakneck speed, taking such random turns, anything seems possible, and the worst of things feels likely. It's the suddenness of it all that makes it so strange. The cancerous surge in nationalism and racism around the world has taken many of us by surprise. I grew up in south-east London, not very far from where black teenager Stephen Lawrence was killed by racist thugs in 1993 while waiting for a bus. His murder left a mark on my generation. When we campaigned against racism, we knew there was a long way to go, but we were hopeful. And for a brief, sunlit moment the world really did seem to be changing. My son was born five years ago, when Barack Obama was President of the United States, and I dreamed then that he might grow up in a better society, perhaps even a post-racial one.

Things ceased to be normal very quickly. In the space of just a few years, far-right and anti-immigrant groups have become visible and powerful across Europe and the United States. In

Poland, nationalists march under the slogan 'Pure Poland, white Poland'. In Italy a right-wing leader rises to popularity on the promise to deport illegal immigrants and turn his back on refugees. White nationalists look to Russia under Vladimir Putin as a defender of 'traditional' values. In German elections in 2017 Alternative für Deutschland wins more than 12 per cent of the vote. Steve Bannon, the former chief strategist to Donald Trump, tells far-right nationalists in France in 2018, 'Let them call you racist, let them call you xenophobes, let them call you nativists. Wear it as a badge of honor.'

While it may be easy to blame white supremacists for this cancer, it's a brand of identity politics that has others in its grip, too. It's infecting people everywhere: Islamic fundamentalists in the Middle East and Pakistan, Hindu nationalists in India, Chinese scholars who turn their back on good science in favour of a worldview that paints the Chinese as having different evolutionary roots from everyone else. They may have different ideas and different histories, but their goal is the same: to assert difference for political gain. This is a twisted ideology that deliberately makes no appeal to a shared humanity, but instead rests on shadowy myths of belonging, on origin stories offering an umbrella to some but not others, sheltering them with false comfort. What nationalism stresses, as the late political scientist Ernst B. Haas wrote, is 'the individual's search for identity with strangers in an impersonal world'.

That desire to belong is powerful, I know. I was raised between cultures, and there's nothing quite so disruptive to your sense of belonging as not fully belonging anywhere, as being brown when everyone else is white in a place that notices these things. But don't be fooled. When they play on those feelings, when they tell you they can return you to a glorious past, offer a community of people just like you, who share your values and your dreams, a common history, they are selling you a myth. Enjoy your culture or religion, have pride in where you live or where your ancestors came from if you like, but don't imagine that these things give you any biological claim. Wear your identity lightly. Don't be sucked into believing that you are so different from others, that your rights have more value, that your blood is a different colour. There is no authenticity except that of personal experience.

The 'race realists', as they call themselves (perhaps because calling yourself a racist is still for now unpalatable even to most racists), work so hard to make the opposite case. They appeal to that dark corner of our souls that wants to believe human difference runs deep, making entire populations special, giving some nations an edge over others. And sadly, this is their moment. Whenever ugly politics become dominant, you can be sure that there are intellectuals and pseudo-intellectuals ready to jump on board. Those with dangerous ideas about 'human nature' and even more dangerous prescriptions for our problems are content to bide their time, knowing that the pendulum will swing their way eventually. Intellectual racism has always existed, and indeed for a chunk of history, it thrived. I believe it is still the toxic little seed at the heart of academia. However dead you might think it is, it needs only a little water, and now it's raining.

That said, what they're doing is also intellectually doomed. I've learned while writing this book that trying to force a biological understanding of race fails, often spectacularly, for the simple reason that it's history that has the answers. Science can't help you here. But then, perhaps the 'race realists' know this. Maybe they know that if we truly want an end to racism, we need to understand the past, to have more equitable education and healthcare, to end discrimination in work and institutions, to be a little more open with our hearts and maybe also with our borders. Maybe they know that the answers are not in our blood, but in us. They are in our actions, in the choices we make, and in the ways in which we treat each other. Maybe their insistent banging of the drum, their increasing violence and anger, is simply to mask the fact that they don't want to make these concessions.

There are plenty of ignorant racists, but the problem is not ignorance alone. The problem is that, even when people know the facts, not all of them actually want an end to racial inequality or even to the idea of race. Some would rather things stayed the way they are, or went backwards. And this means that those committed to the biological reality of race won't back down if the data proves them wrong. There's no incentive for them to admit intellectual defeat. They will simply keep reaching for fresher, more elaborate theories when the old ones fail. If skin colour can't explain racial inequality, then maybe the structure of our brains and bodies will. If not anatomy, then maybe our genes. When then this, too, throws up nothing of value, they will reach for the next thing. All this intellectual jumping through hoops to maintain the status quo. All this to prove what they have always really wanted to know: that they are superior.

Well, keep reaching, keep reaching, keep reaching. One day there will be nothing left to reach for.

Acknowledgements

THIS IS THE BOOK I have wanted to write since I was ten years old, and I have poured my soul into it. I'm grateful that my editors at Fourth Estate and Beacon Press, Louise Haines and Amy Caldwell, didn't hesitate in commissioning it, and that my publicists Michelle Kane and Caitlin Meyer have been my champions throughout. My agents Peter Tallack, Tisse Takagi and Louisa Pritchard were equally supportive. I am very fortunate to have such a loyal, kind-hearted team around me.

I would also like to deeply thank Jon Marks, Eric Turkheimer, Bill Tucker, Jay Kaufman, Subhadra Das, Jennifer Raff, Greg Radick and Billy Griffiths for their generous assistance. My friend, the archaeologist Tim Power, guided me through the British Museum and helped me see the past from a different perspective. I would also like to thank my sister Rima, herself a scholar of race and politics, for her critical feedback, my parents-in-law Neena and Pammi for their kind help at home, and my husband Mukul for his love and patience. When we started dating, Mukul was a fan of the band Fun-Da-Mental and he would often recite the two lines of lyrics I included at the beginning of this book.

In my business, there are real friends – the ones who care – and there's everyone else. My deepest gratitude is reserved for Peter Wrobel, a true friend who, out of the goodness of his heart, scoured the entire manuscript for errors.

This is the second book I have written since my son, Aneurin, was born, and he has made writing both a pleasure. He lights up my days. I don't know what the future holds for him, but I hope he never has to face the struggles that his parents or grandparents did. I hope he understands that how we look, our ancestry, where we come from or where we live are not the most important markers of identity. What makes us who we are is the content of our characters, as Martin Luther King reminded us. And this is entirely in our hands.

Don't forget that, my baby.

References

Prologue

QUOTE AT TOP: Song lyrics by Dennis Webb taken from 'English Breakfast', *Seize the Time* album by Fun-Da-Mental. With deepest thanks to Aki Nawaz and Dennis Webb for their kind permission to reproduce the words.

Factual details about the objects in the British Museum are taken from their labels, with guidance from archaeologist Tim Power.

Bartlett, John. '"Stolen friend": Rapa Nui Seek Return of Moai Statue'. BBC News, 18 November 2018, https://www.bbc.co.uk/news/world-latin-america-46222276 (accessed 19 November 2018)

Blumenbach, Johann Friedrich. *De generis humani varietate native*. Gottingae: Vandenhoek et Ruprecht, 1795

Freedman, Bernard J. 'Caucasian'. *British Medical Journal* (Clinical Research Edition) 288, no. 6418 (3 March 1984), 696–8

Hoock, Holger. 'The British State and the Anglo-French Wars over Antiquities, 1798–1858'. *The Historical Journal* 50, no. 1 (March 2007), 49–72

'Petition: Cultural Genocide: U.S. Government Forces Egyptian Nubians to be Classified as White and Not Black', MoveOn.org, https://petitions.moveon.org/sign/justice for-an-indigenous (accessed 19 November 2018)

'Sir Hans Sloane', British Museum, https://www.britishmuseum. org/about_us/the_museums_story/general_history/sir_hans_ sloane.aspx (accessed 19 November 2018)
Tharoor, Shashi. *Inglorious Empire: What the British Did to India*. London: Hurst & Company, 2017

1 Deep Time

Adcock, Gregory J. et al. 'Mitochondrial DNA Sequences in Ancient Australians: Implications for Modern Human Origins'. *Proceedings of the National Academy of Sciences* 98, no. 2 (January 2001), 537–42
Allen, Harry. 'The Past in the Present? Archaeological Narratives and Aboriginal History', in *Long History, Deep Time: Deepening Histories of Place*, ed. Ann McGrath and Mary Anne Jebb. Australia: ANU Press, 2015, 176–202
Anderson, Kay and Perrin, Colin. '"The Miserablest People in the World": Race, Humanism and the Australian Aborigine'. *Australian Journal of Anthropology* 18, no. 1 (2007), 18–39
Anderson, Kay and Perrin, Colin. 'How Race Became Everything: Australia and Polygenism'. *Ethnic and Racial Studies* 31, no. 5 (2008), 962–90
Athreya, Sheela. 'Picking a Bone with Evolutionary Essentialism', *Anthropology News*, 18 September 2018, http://www. anthropology-news.org/index.php/2018/09/18/picking-a-bone- with-evolutionary-essentialism/ (accessed 15 October 2018)
'Bringing Them Home: Report of the National Inquiry into the Separation of Aboriginal and Torres Strait Islander Children from Their Families', Commonwealth of Australia, 1997
ChangZhu, Jin et al. 'The Homo Sapiens Cave Hominin site of Mulan Mountain, Jiangzhou District, Chongzuo, Guangxi with Emphasis on its Age'. *Chinese Science Bulletin* 54, no. 21 (November 2009), 3848
Clarkson, Chris et al. 'Human Occupation of Northern Australia by 65,000 Years Ago'. *Nature* 547 (20 July 2017), 306–10

'Colonial Frontier Massacres in Central and Eastern Australia 1788–1930', University of Newcastle, Australia, https://c21ch. newcastle.edu.au/colonialmassacres/timeline.php (accessed 7 September 2018)

Dannemann, Michael and Kelso, Janet. 'The Contribution of Neanderthals to Phenotypic Variation in Modern Humans'. *American Journal of Human Genetics* 101, no. 4 (5 October 2017), 578–89

Gibbons, Ann. 'Who Were the Denisovans?' *Science* 333, no. 6046 (26 August 2011), 1084–7

Gosden, Chris. 'Race and Racism in Archaeology: Introduction'. *World Archaeology* 38, no. 1 (2006), 1–7

Gregory, Michael D. et al. 'Neanderthal-Derived Genetic Variation Shapes Modern Human Cranium and Brain'. *Scientific Reports* 7, no. 6308 (December 2017)

Griffiths, Billy. *Deep Time Dreaming*. Carleton, Victoria: Black Inc. Books, 2018

Gunn, R. G. 'Mulka's Cave Aboriginal rock art site: its context and content'. *Records of the Western Australian Museum* 23, 2006, 19–41

Hublin, Jean-Jacques et al. 'New fossils from Jebel Irhoud, Morocco and the Pan-African origin of *Homo sapiens*'. *Nature* 546 (8 June 2017), 289–92

Ingold, Tim. 'Beyond Biology and Culture: The Meaning of Evolution in a Relational World'. *Social Anthropology* 12, no. 2 (June 2004), 209–21

Kaplan, Sarah. 'Humans Didn't Outsmart the Neanderthals. We Just Outlasted Them', *Washington Post*, 1 November 2017. www.washingtonpost.com/news/speaking-of-science/wp/2017/ 11/01/humans-didnt-outsmart-the-neanderthals-we-just-out lasted-them/?utm_term=.4a78010999de (accessed 16 October 2018)

Karakostis, Fotios Alexandros et al. 'Evidence for Precision Grasping in Neandertal Daily Activities'. *Science Advances* 4, no. 9 (26 September 2018)

Kolbert, Elizabeth, 'Our Neanderthals, Ourselves', *New Yorker*, 12 February 2015, https://www.newyorker.com/news/daily-comment/neanderthals (accessed 16 October 2018)

Krause, Johannes et al. 'The Complete Mitochondrial DNA Genome of an Unknown Hominin from Southern Siberia'. *Nature* 464 (8 April 2010), 894–7

Marris, Emma. 'News: Neanderthal Artists Made Oldest-known Cave Paintings'. *Nature* (22 February 2018)

Milks, Annemieke. 'We Haven't Been Giving Neanderthals Enough Credit.' *Popular Science*, 26 June 2018, www.popsci.com/neanderthal-hunting-spears (accessed 16 October 2018)

Mills, Charles W. *The Racial Contract*. Ithaca, NY: Cornell University Press, 1997

Mooallem, Jon, 'Neanderthals Were People, Too'. *New York Times*, 11 January 2017, https://www.nytimes.com/2017/01/11/magazine/neanderthals-were-people-too.html (accessed 16 October 2018)

'Neanderthals Thought Like We Do', Max-Planck-Gesellschaft, 22 February 2018, https://www.mpg.de/11948095/neandertals-cave-art (accessed 11 October 2018)

'Neanderthals Were Too Smart for Their Own Good'. *The Telegraph*, 18 November 2011, www.telegraph.co.uk/news/science/science-news/8898321/Neanderthals-were-too-smart-for-their-own-good.html (accessed 16 October 2018)

Noonuccal, Oodgeroo. 'Stone Age'. In *The Dawn Is at Hand*, London: Marion Boyars, 1992

Parsons, Meg. 'Creating a Hygienic Dorm: The Refashioning of Aboriginal Women and Children and the Politics of Racial Classification in Queensland 1920s–40s'. *Health and History* 14, no. 2 (2012), 112–39

Pascoe, Bruce. *Dark Emu, Black Seeds: Agriculture or Accident*. Broome, Western Australia: Magabala Books, 2014

Porr, Martin. 'Essential Questions: Modern Humans and the Capacity for Modernity', in *Southern Asia, Australia and the Search for Human Origins*, ed. Robin Dennell and Martin Porr. Cambridge: Cambridge University Press, 2014, 257–64

Porr, Martin and Matthews, Jacqueline M. 'Post-colonialism, Human Origins and the Paradox of Modernity. *Antiquity* 91, no. 358 (August 2017), 1058–68

Prüfer, Kay et al. 'The Complete Genome Sequence of a Neanderthal from the Altai Mountains'. *Nature* 505 (2 January 2014), 43–9

Qiu, Jane. 'The Forgotten Continent'. *Nature* 535 (14 July 2016), 218–20

Quach, Hélène et al. 'Genetic Adaptation and Neandertal Admixture Shaped the Immune System of Human Populations'. *Cell* 167, no. 3 (20 October 2016), 643–56

Scerri, Eleanor M.L. et al. 'Did Our Species Evolve in Subdivided Populations across Africa, and Why Does It Matter?' *Trends in Ecology & Evolution* 33, no. 8 (August 2018), 582–94

Thorne, Alan G. and Wolpoff, Milford H. 'Regional Continuity in Australasian Pleistocene Hominid Evolution'. *American Journal of Physical Anthropology* 55, no. 3 (July 1981), 337–49

Thorne, Alan G. and Wolpoff, Milford H. 'The Multiregional Evolution of Humans'. *Scientific American* 266, no. 4 (April 1992), 76–83

van der Kaars, Sander et al. 'Humans Rather than Climate the Primary Cause of Pleistocene Megafaunal Extinction in Australia'. *Nature Communications* 8, no. 14142 (20 January 2017)

Webb, Steve. *Made in Africa: Hominin Explorations and the Australian Skeletal Evidence*. Cambridge, MA: Academic Press, 2018

Wenban-Smith, Frances, 'Neanderthals Were No Brutes – Research Reveals they May Have Been Precision Workers', *The Conversation*, 26 September 2018, https://theconversation. com/neanderthals-were-no-brutes-research-reveals-they-may-have-been-precision-workers-103858?utm_medium=Social& utm_source=Twitter#Echobox-1538043704 (accessed 11 October 2018)

2 It's a Small World

'The African-American Mosaic', Library of Congress, http://www.loc.gov/exhibits/african/afam002.html (accessed 12 October 2017)

Berry, Daina Ramey. *The Price for Their Pound of Flesh: The Value of the Enslaved, from Womb to Grave, in the Building of a Nation*. Boston, MA: Beacon Press, 2017

Cartwright, Samuel A. 'Diseases and Peculiarities of the Negro Race'. *De Bow's Review*, Southern and Western States 11, New Orleans, 1851

Desmond, Adrian and Moore, James. *Darwin's Sacred Cause: How a Hatred of Slavery Shaped Darwin's Views on Human Evolution*. New York: Houghton Mifflin Harcourt, 2009

Douglas, Bronwen. 'Climate to Crania: Science and the Racialization of Human Difference', in *Foreign Bodies: Oceania and the Science of Race 1750–1940*, ed. Bronwen Douglas and Chris Ballard. Canberra: ANU Press, 2008

Fields, Karen E. and Fields, Barbara J. *Racecraft: The Soul of Inequality in American Life*, Reprint edn. London and New York: Verso Books, 2014

Geertz, Clifford. *The Interpretation of Cultures*. New York: Basic Books, 1973

Gobineau, Arthur, comte de. *The Inequality of Human Races*. 1853; English translation by Adrian Collins. London: William Heinemann, 1915

Gould, Stephen Jay. *The Mismeasure of Man*. New York: W. W. Norton, 1981

'The Hottentot Venus Is Going Home'. *Journal of Blacks in Higher Education*, no. 35 (Spring 2002), 63

Huxley, Thomas Henry. *Collected Essays: Volume 3, Science and Education*. London: Macmillan, 1893

Keller, Mitch. 'The Scandal at the Zoo'. *New York Times*, 6 August 2006

Lewis, Bernard. 'The Historical Roots of Racism'. *The American Scholar* 67, no. 1 (Winter 1998), 17–25

Malik, Kenan. *Strange Fruit: Why Both Sides are Wrong in the Race Debate*. London: Oneworld, 2008

Marks, Jonathan. *Is Science Racist?* Cambridge: Polity Press, 2017

O'Brien, Conor Cruise. 'Thomas Jefferson: Radical and Racist'. *The Atlantic*, October 1996

Radick, Gregory. 'Darwin and Humans', in *The Cambridge Encyclopedia of Darwin and Evolutionary Thought*, ed. Michael Ruse. Cambridge: Cambridge University Press, 2013, 173–81

Radick, Gregory. 'How and Why Darwin Got Emotional about Race', in *Historicizing Humans: Deep Time, Evolution and Race in Nineteenth Century British Sciences*, ed. Efram Sera-Shriar. Pittsburgh: University of Pittsburgh Press, 2018

Solly, S., Moojen. Geo and Lindfors, Bernth. 'Courting the Hottentot Venus'. *Africa: Rivista trimestrale di studi e documentazione dell'Istituto italiano per l'Africa e l'Oriente*. Vol. 40, no. 1 (March 1985), 133–48

Swanton, John R. 'Review: The Inequality of Human Races by Arthur De Gobineau and Adrian Collins'. *American Anthropologist*, New Series 18, no. 3 (July–September 1916), 429–31

Zeitoun, Charline. 'In the Days of Human Zoos'. CNRS, 22 November 2016, https://news.cnrs.fr/articles/in-the-days-of-human-zoos (accessed 9 October 2017)

3 Scientific Priestcraft

A. B. S. 'Reviewed Work: The Passing of the Great Race: Or the Racial Basis of European History by Madison Grant'. *American Historical Review* 22, no. 4 (July 1917), 842–4

Black, Edwin. *IBM and the Holocaust: The Strategic Alliance between Nazi Germany and America's Most Powerful Corporation*. New York: Crown, 2001

Boulter, Michael. 'The Rise of Eugenics, 1901–14', in
  Bloomsbury Scientists: Science and Art in the Wake of Darwin.
  London: UCL Press, 2017, 102–14

Burke, Jason and Oltermann, Philip. 'Germany moves to atone
  for "forgotten genocide" in Namibia'. *The Guardian*, 25
  December 2016

Chung, Yuehtsen Juliette. 'Better Science and Better Race?:
  Social Darwinism and Chinese Eugenics'. *Isis* 105, no. 4
  (December 2014), 793–802

Durant, John R. 'Scientific Naturalism and Social Reform in the
  Thought of Alfred Russel Wallace'. *British Journal for the
  History of Science* 12, no. 1 (March 1979), 31–58

'The Eugenics Record Office', *Science* 37, no. 954 (11 April
  1913), 553–4

'First International Eugenics Congress, London, July 24th to
  July 30th, 1912, University of London, South Kensington:
  Programme and time table', https://archive.org/stream/b2243
  9833/b22439833_djvu.txt (accessed 12 January 2018)

Gabriel, Elliott. 'National Hygiene and "Inferior Offspring":
  Japan's Eugenics Victims Demand Justice'. *Mint Press News*,
  19 March 2018, https://www.mintpressnews.com/inferior-
  offspring-japans-eugenics-victims-demand-justice/239078/
  (accessed 26 March 2018)

Galton, Francis. 'Hereditary Character and Talent'. *Macmillan's
  Magazine* 12 (1865), 157–66

Grant, Madison. *The Passing of the Great Race: Or the Racial Basis
  of European History*. New York: Charles Scribner's Sons, 1916

Hall, Prescott F. 'Selection of Immigration'. *Annals of the
  American Academy of Political and Social Science* 24 (July
  1904), 169–84

Hawes, James. *The Shortest History of Germany*. London: Old
  Street, 2017

Heim, Susanne, Sachse, Carola and Walker, Mark. *The Kaiser
  Wilhelm Society under National Socialism*. Cambridge:
  Cambridge University Press, 2009

'History of the Kaiser Wilhelm Society under National Socialism',
    Max Planck Society, https://www.mpg.de/9811513/kws-under-
    national-socialism (accessed 22 October 2017)
'International Eugenics Congress', *The Scientific Monthly* 12,
    no. 4 (April 1921), 383–4
Johnson, Corey. 'California Was Sterilising its Female Prisoners
    as Late as 2010'. *The Guardian*, 8 November 2013
Johnson, Roswell H. 'Eugenics and So-Called Eugenics'.
    American Journal of Sociology 20, no. 1 (July 1914), 98–103
Kohn, Marek. *The Race Gallery: The Return of Racial Science*.
    London: Jonathan Cape, 1995
Lee, Jonathan H. X. *History of Asian Americans: Exploring
    Diverse Roots*. Santa Barbara, CA: Greenwood, 2015
Lewontin, Richard C. 'Biological Determinism'. *Tanner Lectures
    on Human Values* 4 (1983), 147–83
Mukherjee, Siddhartha. *The Gene: An Intimate History*. New
    York: Scribner, 2016
Müller-Hill, Benno. 'The Blood from Auschwitz and the Silence
    of the Scholars'. *History and Philosophy of the Life Sciences*
    21, no. 3 (1999), 331–65
Priemel, Kim Christian. 'Review: The Kaiser Wilhelm Society
    under National Socialism by Susanne Heim, Carola Sachse
    and Mark Walker'. *Journal of Modern History* 83, no. 1
    (March 2011), 216–18
Schaffer, Gavin. '"Like a Baby with a Box of Matches": British
    Scientists and the Concept of "Race" in the Inter-War Period'.
    British Journal for the History of Science 38, no. 3 (September
    2005), 307–24
Sussman, Robert Wald. *The Myth of Race: The Troubling
    Persistence of an Unscientific Idea*. Cambridge, MA and
    London: Harvard University Press, 2014
Tuffs, Annette. 'German Research Society Apologises to Victims
    of Nazis'. *British Medical Journal* (16 June 2001), 322
Weigmann, Katrin. 'In the Name of Science'. *EMBO Reports* 2,
    no. 10 (2001), 871–5

Whitman, James Q. *Hitler's American Model: The United States and the Making of Nazi Race Law*. Princeton, NJ and Oxford: Princeton University Press, 2017

4 Inside the Fold

With thanks to King's College London for access to their special collections.

Barash, David. 'Review: Race, Evolution and Behavior'. *Animal Behaviour* 49, no. 4 (April 1995), 1131–3

Campbell, Michael C. and Tishkoff, Sarah A. 'African Genetic Diversity: Implications for Human Demographic History, Modern Human Origins, and Complex Disease Mapping'. *Annual Review of Genomics and Human Genetics* 9 (2008), 403–33

Dutton, Edward. 'Obituary: Tatu Vanhanen 1929–2015'. *Mankind Quarterly* 56, no. 2 (2015), 225–32

Edwards, A. W. F. 'Human Genetic Diversity: Lewontin's Fallacy'. *Bioessays* 25, no. 8 (August 2003), 798–801

Garrett, Henry E. 'Klineberg's Chapter on Race and Psychology'. *Mankind Quarterly* (July 1960), 15–22

Garrett, Henry E. 'The Equalitarian Dogma'. *Mankind Quarterly* (April 1961), 253–7

Gayre, R. 'The Dilemma of Inter-Racial Relations'. *Mankind Quarterly* 6, no. 4 (April–June 1966)

Genoves, Santiago. 'Racism and "The Mankind Quarterly"'. *Science* 134, no. 3493 (8 December 1961), 1928–32

Harrison, Geoffrey Ainsworth. 'Reviewed Work: The Mankind Quarterly by R. Gayre'. *Man* 61 (September 1961), 163–4

International Consortium of Investigative Journalists Offshore Leaks Database, https://offshoreleaks.icij.org/nodes/21000166 (accessed 19 December 2017)

Kelsey, Tim. 'Ulster University Took Grant from Fund Backing Whites'. *The Independent*, 9 January 1994

Kunzelman, Michael. 'University Accepted $458K from Eugenics

Fund'. Associated Press, 25 August 2018, https://apnews.com/
a9791e6174374437b3bbe17af8b76215 (accessed 27 August
2018)

Lewontin, Richard Charles. 'The Apportionment of Human
Diversity'. *Evolutionary Biology* 6 (1972) 381–98

Lipphardt, Veronika. 'From "Races" to "Isolates" and
"Endogamous Communities": Human Genetics and the
Notion of Human Diversity in the 1950s', in *Human Heredity
in the Twentieth Century*, ed. Bernd Gausemeier, Staffan
Müller-Wille and Edmund Ramsden. London: Pickering &
Chatto, 2013, 55–68

Miller, Adam. 'The Pioneer Fund: Bankrolling the Professors of
Hate'. *Journal of Blacks in Higher Education*, no. 6 (Winter
1994–5), 58–61 (a version of this article originally appeared in
the *Los Angeles Times* in 1994)

Montagu, Ashley. *Man's Most Dangerous Myth: The Fallacy of
Race*. New York: Columbia University Press, 1942

Montagu, Ashley. 'The Genetical Theory of Race, and
Anthropological Method'. *American Anthropologist* 44, no. 3
(July–September 1942)

Radick, Gregory. 'Beyond the "Mendel-Fisher Controversy"'.
Science 350, no. 6257 (9 October 2015), 159–60

Radick, Gregory. 'Presidential Address: Experimenting with the
Scientific Past'. *British Journal for the History of Science* 49,
no. 2 (June 2016), 153–72

'Reginald Ruggles Gates Collection', King's College London,
https://www.kcl.ac.uk/library/archivespec/special-collections/
Individualcollections/rugglesgates.aspx (accessed 2 November
2017)

Rosenberg, Noah A. et al. 'Genetic Structure of Human
Populations'. *Science* 298, no. 5602 (20 December 2002), 2381–5

Schaffer, Gavin. '"Scientific" Racism Again?: Reginald Gates, the
"Mankind Quarterly" and the Question of "Race" in Science
after the Second World War'. *Journal of American Studies* 41,
no. 2 (August 2007), 253–78

Selcer, Perrin. 'Beyond the Cephalic Index: Negotiating Politics to Produce UNESCO's Scientific Statements on Race'. *Current Anthropology* 53, Supplement 5 (April 2012)

Silverman, Rachel. 'The Blood Group "Fad" in Post-War Racial Anthropology'. *Kroeber Anthropological Society Papers* 84 (2000), 11–27

Singh, Indera P. and Singh, Darshan. 'The Study of ABO Blood Groups of Sainis of Punjab'. *American Journal of Physical Anthropology* 19, no. 3 (September 1961), 223–6

Škerlj, Božo. 'The Mankind Quarterly'. Man 60 (November 1960), 172–3

Sussman, Robert Wald. *The Myth of Race: The Troubling Persistence of an Unscientific Idea.* Cambridge, MA and London: Harvard University Press, 2014

Tucker, William H. *The Funding of Scientific Racism: Wickliffe Draper and the Pioneer Fund.* Urbana: University of Illinois Press, 2002

UNESCO, 'The Race Concept: Results of an Inquiry', in *The Race Question in Modern Science.* Paris: UNESCO, 1952

Witherspoon, D. J. et al. 'Genetic Similarities Within and Between Human Populations'. *Genetics* 176, no. 1 (May 2007), 351–9

5 Race Realists

Burleigh, Nina. 'Steve Bannon, Jared Taylor and the Radical Right's Ivy League Pedigree'. *Newsweek*, 23 March 2017, http://www.newsweek.com/bannon-spencer-trump-alt-right-breitbart-infowars-yale-gottfried-oathkeepers-572585 (accessed 16 November 2017)

'A Convocation of Bigots: The 1998 American Renaissance Conference'. *Journal of Blacks in Higher Education*, no. 21 (Autumn 1998), 120–4

'Intelligence Editorial Board', Elsevier, https://www.journals. elsevier.com/intelligence/editorial-board (accessed 10 November 2017 and 20 November 2018)

Jensen, Arthur. 'How Much Can We Boost IQ and Scholastic Achievement?' *Harvard Educational Review* 39, no. 1 (April 1969), 1–123

Kelsey, Tim, and Rowe, Trevor. 'Academics "were funded by racist American trust"'. *Independent on Sunday*, 4 March 1990

Lane, Charles. 'The Tainted Sources of "The Bell Curve"'. *New York Review of Books*, 1 December 1994

Levin, Michael. Letter to the Editor: 'Howard Beach Turns a Beam on Racial Tensions'. *New York Times*, 11 January 1987

Lynn, Richard. 'The Intelligence of American Jews'. *Personality and Individual Differences* 36, no. 1 (January 2004), 201–6

Lynn, Richard and Meisenberg, Gerhard. Review article: 'The Average IQ of Sub-Saharan Africans: Comments on Wicherts, Dolan, and van der Maas'. *Intelligence* 38, no. 1 (January–February 2010), 21–9

Mehler, Barry. 'The New Eugenics: Academic Racism in the U.S. Today'. *Science for the People* 15, no. 3 (May–June 1983), 18–23

Mehler, Barry. 'Rightist on the Rights Panel'. *The Nation* (7 May 1988), 640–1

'Michael Levin', Southern Poverty Law Center, https://www.splcenter.org/fighting-hate/extremist-files/individual/michael-levin (accessed 13 November 2017)

Murray, Charles A. and Herrnstein, Richard. *The Bell Curve: Intelligence and Class Structure in American Life.* 1994; New York: Free Press Paperbacks, 1996

Nelson, Louis. 'Clinton Ad Ties Trump to KKK, White Supremacists'. POLITICO, 25 August 2016, https://www.politico.com/story/2016/08/clinton-ad-kkk-trump-227404 (accessed 28 November 2017)

Pearson, Roger. 'Immigration into Britain'. *Northlander* 1, no. 1 (April 1958), 2

Rosenthal, Steven J. 'The Pioneer Fund: Financier of Fascist Research'. *American Behavioral Scientist* 39, no. 1 (September–October 1995), 44–61

Santiago, Frank. 'Rights Official Has Racial "Purity" Links'. *Des Moines Register* (28 February 1988)

Schulson, Michael. 'Race, Science, and Razib Khan'. Undark, 28 February 2017, https://undark.org/article/race-science-razib-khan-racism/ (accessed 20 November 2018)

Scott, Ralph. 'Arthur Jensen: A Latter-Day "Enemy of the People"?' *Mankind Quarterly* 53, no. 3–4 (2013)

Siegel, Eric. 'The Real Problem with Charles Murray and "The Bell Curve"'. *Scientific American*, https://blogs.scientific american.com/voices/the-real-problem-with-charles-murray-and-the-bell-curve/, 12 April 2017 (accessed 6 November 2017)

Stein, Joel. 'Milo Yiannopoulos is the Pretty, Monstrous Face of the Alt-Right'. Bloomberg, 15 September 2016, https://www.bloomberg.com/features/2016-america-divided/milo-yiannopoulos/ (accessed 20 November 2018)

Stolarski, Maciej, Zajenkowski, Marcin and Meisenberg, Gerhard. 'National Intelligence and Personality: Their Relationships and Impact on National Economic Success'. *Intelligence* 41, no. 2 (March–April 2013), 94–101

Sussman, Robert Wald. 'America's Virulent Racists: The Sick Ideas and Perverted "science" of the American Renaissance Foundation'. *Salon*, 11 October 2014, https://www.salon.com/2014/10/11/americas_virulent_racists_the_sick_ideas_and_perverted_science_of_the_american_renaissance_foundation/ (accessed 13 November 2017)

van der Merwe, Ben. 'Exposed: London's Eugenics Conference and its Neo-Nazi Links. *London Student*, 10 January 2018, http://londonstudent.coop/news/2018/01/10/exposed-london-eugenics-conferences-neo-nazi-links/amp/?__twitter_impression=true (accessed 15 January 2018)

Weyher, Harry F. 'The Pioneer Fund, the Behavioral Sciences, and the Media's False Stories'. *Intelligence* 26, no. 4 (1998), 319–36

6 Human Biodiversity

'Ancestry Names Margo Georgiadis Chief Executive Officer',
Ancestry.com, 19 April 2018, https://www.ancestry.com/
corporate/newsroom/press-releases/ancestry-names-margo-
georgiadis-chief-executive-officer (accessed 20 November
2018)

Boodman, Eric. 'White Nationalists Are Flocking to Genetic
Ancestry Tests. Some Don't Like what they Find'. *STAT News*,
16 August 2017, https://www.statnews.com/2017/08/16/
white-nationalists-genetic-ancestry-test/ (accessed 7 March
2018)

Cavalli-Sforza, Luigi Luca. *Genes, Peoples, and Languages.*
London: Penguin, 2001

Cavalli-Sforza, Luca. 'The Human Genome Diversity Project:
An Address Delivered to a Special Meeting of UNESCO',
Paris (12 September 1994)

Evans, Rob. 'Military Scientists Tested Mustard Gas on Indians'.
The Guardian, 1 September 2007, https://www.theguardian.
com/uk/2007/sep/01/india.military (accessed 3 August 2018)

Gannett, Lisa. 'Racism and Human Genome Diversity Research:
The Ethical Limits of "Population Thinking"'. *Philosophy of
Science* 68, no. 3 (2001), Supplement S479–S492

Gates, Henry Louis. *Finding Oprah's Roots: Finding Your Own.*
New York: Crown, 2007

Kennett, Debbie A. et al. 'The Rise and Fall of BritainsDNA:
A Tale of Misleading Claims, Media Manipulation and
Threats to Academic Freedom'. *Genealogy* 2, no. 4 (2
November 2018)

Lipphardt, Veronika. '"Geographical Distribution Patterns of
Various Genes": Genetic Studies of Human Variation after
1945'. *Studies in History and Philosophy of Biological and
Biomedical Sciences* 47 (September 2014), 50–61

MacDougald, Park and Willick, Jason. 'The Man Who Invented
Identity Politics for the New Right'. *New York*, 30 April 2017,

http://nymag.com/daily/intelligencer/2017/04/steve-sailer-invented-identity-politics-for-the-alt-right.html (accessed 4 January 2018)

Malik, Kenan. *Strange Fruit: Why Both Sides Are Wrong in the Race Debate*. Oxford: Oneworld, 2008

Marks, Jonathan. *Human Biodiversity: Genes, Race, and History*. New York: Aldine de Gruyter, 1995

Murray, Charles A. and Herrnstein, Richard. *The Bell Curve: Intelligence and Class Structure in American Life*. 1994; New York: Free Press, 1996

Olson, Steve. 'The Genetic Archaeology of Race'. *The Atlantic*, April 2001, https://www.theatlantic.com/magazine/archive/2001/04/the-genetic-archaeology-of-race/302180/ (accessed 14 February 2018)

'Oprah Winfrey's Surprising DNA Test', Ancestry.com, https://blogs.ancestry.com/cm/the-surprising-facts-oprah-winfrey-learned-about-her-dna/ (accessed 7 March 2018)

Radin, Joanna. 'Human Genome Diversity Project: History', in *International Encyclopedia of the Social & Behavioral Sciences*, 2nd edn, vol. 11, ed. James D. Wright. Oxford: Elsevier, 2015, 306–10

Reardon, Jenny. *Race to the Finish: Identity and Governance in an Age of Genomics*. Princeton, NJ: Princeton University Press, 2004

Reardon, Jenny, 'Finding Oprah's Roots, Losing the World: Beyond the Liberal AntiRacist Genome'. Unpublished paper presented at Berkeley Workshop on Environmental Politics, 23 October 2009

Taubes, Gary. 'Scientists Attacked for "Patenting" Pacific Tribe'. *Science* 270, no. 5239 (17 November 1995), 1112

UNESCO IBC Working Group on Population Genetics. *Draft Report: Bioethics and Human Population Genetics Research*. Third Session of the IBC, 27–29 September 1995

Walker, Matthew. 'First, Do Harm'. *Nature* 482 (9 February 2012), 148–52

WHO Scientific Group on Research in Population Genetics of
Primitive Groups, *Research in population genetics of primitive
groups: Report of a WHO Scientific Group [meeting held in
Geneva from 27 November to 3 December 1962]*. Geneva:
World Health Organization, 1964

7 Roots

Allentoft, Morten E. et al. 'Population Genomics of Bronze Age
Eurasia'. *Nature* 522, no. 7555 (June 2015), 167–72
Bracc, Selina et al. 'Population Replacement in Early Neolithic
Britain'. *bioRxiv*, 18 February 2018, https://www.biorxiv.org/
content/early/2018/02/18/267443
Crawford, Nicholas G. et al. 'Loci associated with skin
pigmentation identified in African populations'. *Science*,
published online 12 October 2017
Devega, Chauncey. 'Cheddar Man is "black"! Another racial
panic for white supremacists'. *Salon*, 12 February 2018, https://
www.salon.com/amp/cheddar-man-is-black-another-racial-
panic-for-white-supremacists (accessed 20 November 2018)
Hunt-Grubbe, Charlotte. 'The Elementary DNA of Dr Watson'.
Sunday Times, 14 October 2007
Kristiansen, Kristian et al. 'Re-theorising Mobility and the
Formation of Culture and Language among the Corded Ware
Culture in Europe'. *Antiquity* 91, no. 356 (April 2017),
334–47
Long, Tegwen et al. 'Cannabis in Eurasia: Origin of Human Use
and Bronze Age Trans-continental Connections'. *Vegetation
History and Archaeobotany* 26, no. 2 (March 2017), 245–58
Mathieson, Iain et al. 'Genome-wide Patterns of Selection in 230
Ancient Eurasians'. *Nature* 528, no. 7583 (December 2015),
499–503
Mathieson, Iain et al. 'The Genomic History of Southeastern
Europe. *Nature*, 21 February 2018, https://www.nature.com/
articles/nature25778

Moorjani, Priya et al. 'Genetic Evidence for Recent Population
    Mixture in India'. *American Journal of Human Genetics* 93,
    no. 3 (September 2013), 422–38
'Morrissey: Big mouth strikes again', *New Musical Express*,
    December 2007
Narasimhan, Vagheesh M. et al. 'The Genomic Formation of
    South and Central Asia'. *bioRxiv*, 31 March 2018, https://
    www.biorxiv.org/content/early/2018/03/31/292581
'Nobelist Jim Watson is honored on his 90th birthday', Cold
    Spring Harbour Laboratory website, 11 April 2018, https://
    www.cshl.edu/nobelist-jim-watson-honored-90th-birthday/
    (accessed 18 April 2018)
Olalde, Iñigo et al. 'Derived Immune and Ancestral Pigmentation
    Alleles in a 7,000-year-old Mesolithic European'. *Nature* 507,
    no. 7491 (13 March 2014), 225–8
Olalde, Iñigo et al. 'The Beaker Phenomenon and the Genomic
    Transformation of Northwest Europe'. *Nature* 555 (8 March
    2018), 190–6
Pasha-Robinson, Lucy. 'One in Three Black, Asian or Minority
    Ethnic People Racially Abused Since Brexit, Study Reveals'.
    The Independent, 17 March 2017, http://www.independent.
    co.uk/news/uk/home-news/one-three-black-asian-minority-
    ethnic-bame-racism-abuse-assault-brexit-hate-crime-tuc-study-
    a7634231.html (accessed 6 March 2018)
Reade, Brian. 'Dark-skinned Cheddar Man is Hard Cheese for
    the Racist Morons of the far right'. *Mirror*, 10 February 2018,
    https://www.mirror.co.uk/news/uk-news/dark-skinned-cheddar-
    man-hard-11999683 (accessed 20 November 2018)
Reich, David. *Who We Are and How We Got Here*. Oxford:
    Oxford University Press, 2018
Rutherford, Adam. *A Brief History of Everyone Who Ever Lived:
    The Stories in Our Genes*. London: Weidenfeld and Nicolson,
    2016
University College London press release, 'Face of first Brit
    revealed', 7 February 2018, https://www.ucl.ac.uk/news/news-

articles/0218/070218-Face-of-cheddar-man-revealed (accessed 10 March 2018)

Watson, James. *The Double Helix: A Personal Account of the Discovery of the Structure of DNA*. New York: Atheneum, 1968

8 Origin Stories

Arnold, Bettina. "The Past as Propaganda: Totalitarian Archaeology in Nazi Germany'. *Antiquity* 64, No 244 (September 1990), 464–78

Arnold, Bettina. "'Arierdämmerung": Race and Archaeology in Nazi Germany'. *World Archaeology* 38, no. 1 (March 2006), 8–31

'Asifa Bano: The child rape and murder that has Kashmir on edge', BBC News website, 12 April 2018, http://www.bbc.co.uk/news/world-asia-india-43722714 (accessed 26 April 2018)

Bardill, Jessica et al. 'Advancing the Ethics of Paleogenomics'. *Science* 360, no. 6387 (27 April 2018), 384–5

Colavito, Jason, 'White Nationalists and the Solutrean Hypothesis', blog, 31 January 2014, http://www.jasoncolavito.com/blog/white-nationalists-and-the-solutrean-hypothesis (accessed 22 March 2018)

Crossland, David. 'Germany Recalls Myth That Created the Nation'. *Spiegel Online* (28 August 2009), http://www.spiegel.de/international/germany/battle-of-the-teutoburg-forest-germany-recalls-myth-that-created-the-nation-a-644913.html (accessed 28 April 2018)

'Editorial: Use and abuse of ancient DNA'. *Nature* 555 (March 2018), 559

Friese, Kai. '4500-year-old DNA from Rakhigarhi Reveals Evidence that Will Unsettle Hindutva Nationalists'. *India Today*, 31 August 2018, https://www.indiatoday.in/amp/magazine/cover-story/story/20180910-rakhigarhi-dna-study-findings-indus-valley-civilisation-1327247-2018-08-31 (accessed 5 September 2018)

Ghose, Tia. 'Ancient Kennewick Man Finally Laid to Rest'.
    Live Science, 22 February 2017, https://www.livescience.
    com/57977-kennewick-man-reburied.html (accessed 20
    March 2018)
Gibbons, Ann. 'There's No Such Thing as a "Pure" European –
    or Anyone Else'. *Science* (15 May 2017)
Green, Sara Jean. '"A Wrong had Finally Been Righted": Tribes
    Bury Remains of Ancient Ancestor Known as Kennewick
    Man'. *Seattle Times*, 19 February 2017
Harmon, Amy. 'Indian Tribe Wins Fight to Limit Research of its
    DNA'. *New York Times*, 21 April 2010
Harmon, Amy. 'Geneticists See Work Distorted for Racist Ends'.
    New York Times, 18 October 2018
Jain, Rupam and Lasseter, Tom. 'Special Report – By Rewriting
    History, Hindu Nationalists Aim to Assert their Dominance
    over India'. Reuters, 6 March 2018, https://www.reuters.com/
    investigates/special-report/india-modi-culture/ (accessed 23
    April 2018)
Jha, D. N. 'Against Communalising History'. *Social Scientist* 26,
    no. 9/10 (September–October 1998), 52–62
Lindo, John et al. 'Ancient Individuals from the North American
    Northwest Coast Reveal 10,000 Years of Regional Genetic
    Continuity'. *Proceedings of the National Academy of Sciences
    of the United States of America* 114, no. 16 (April 2017),
    4093–8
Mees, Bernard. 'Hitler and Germanentum'. *Journal of
    Contemporary History* 39, no. 2 (April 2004), 255–70
'Nephi's Neighbors: Book of Mormon Peoples and Pre-
    Columbian Populations', Fair Mormon website, https://www.
    fairmormon.org/conference/august-2003/nephis-neighbors-
    book-of-mormon-peoples-and-pre-columbian-populations
    (accessed 23 March 2018)
O'Brien, Michael J. et al. 'On Thin Ice: Problems with Stanford
    and Bradley's Proposed Solutrean Colonisation of North
    America'. *Antiquity* 88, no. 340 (May 2014), 606–13

O'Brien, Michael J. et al. 'Solutreanism', *Antiquity* 88, no. 340 (January 2015), 622–4

O'Fallon, Brendan D. and Fehren-Schmitz, Lars. 'Native Americans Experienced a Strong Population Bottleneck Coincident with European Contact.' *Proceedings of the National Academy of Sciences*, Vol. 108, no. 51 (December 2011), 20444–8

Oppenheimer, Stephen Bradley, Bruce and Stanford, Dennis. 'Solutrean Hypothesis: Genetics, the Mammoth in the Room'. *World Archaeology* 46, no. 5 (October 2014), 752–74

Pillalamarri, Akhilesh. 'When History Gets Political: India's Grand "Aryan" Debate and the Indus Valley Civilization'. *The Diplomat*, 18 August 2016, https://thediplomat.com/ 2016/08/ when-history-gets-political-indias-grand-aryan-debate-and-the-indus-valley-civilization/ (accessed 29 April 2018)

Preston, Douglas. 'The Kennewick Man Finally Freed to Share His Secrets'. *Smithsonian Magazine*, September 2014

Raff, Jennifer. 'Rejecting the Solutrean Hypothesis'. *The Guardian*, 21 February 2018, https://www.theguardian.com/science/2018/feb/21/rejecting-the-solutrean-hypothesis-the-first-peoples-in-the-americas-were-not-from-europe (accessed 19 March 2018)

Raff, Jennifer A. and Bolnick, Deborah A. 'Does Mitochondrial Haplogroup X Indicate Ancient Trans-Atlantic Migration to the Americas? A Critical Re-Evaluation'. *PaleoAmerica* 1, no. 4 (November 2015), 297–304

Raghavan, Maanasa et al, 'Upper Palaeolithic Siberian Genome Reveals Dual Ancestry of Native Americans'. *Nature* 505 (January 2014), 87–91

Rasmussen, Morten et al. The Genome of a Late Pleistocene Human from a Clovis Burial Site in Western Montana. *Nature* 506 (February 2014), 225–9

Sharma, Manimugdha S. 'Faking History Starts Online'. *The Times of India*, 19 November 2017, https://timesofindia.india times.com/home/sunday-times/faking-history-starts-online/ articleshow/61705453.cms (accessed 29 April 2018)

Srivastava, Sushil. 'The Abuse of History: A Study of the White
    Papers on Ayodhya'. *Social Scientist* 22, no. 5/6 (May–June
    1994), 39–51
Stanford, Dennis and Bradley, Bruce. 'Reply to O'Brien et al'.
    Antiquity 88, no. 340, (January 2015), 614–21
TallBear, Kim. *Native American DNA: Tribal Belonging and the
    False Promise of Genetic Science.* Minneapolis: University of
    Minnesota Press, 2013
TallBear, Kim. 'Genomic Articulations of Indigeneity'. *Social
    Studies of Science* 43, no. 4 (August 2013), 509–33
Thapar, Romila. *The Past as Present.* New Delhi: Aleph, 2014
Thapar, Romila. 'Can Genetics Help Us Understand Indian
    Social History?' *Cold Spring Harbor Perspectives in Biology* 6,
    no. 11 (November 2014)
Wu, Qinglong et al. 'Outburst Flood at 1920 BCE Supports
    Historicity of China's Great Flood and the Xia dynasty'.
    Science 353, no. 6299 (August 2016), 579–82
Zimmer, Carl. 'New Study Links Kennewick Man to Native
    Americans'. *New York Times*, 19 June 2015

9 Caste

Bhavnani, Rikhil R. and Lee, Alexander. 'Does Affirmative Action
    Worsen Bureaucratic Performance? Evidence from the Indian
    Administrative Service'. April 2018, https://faculty.polisci.wisc.
    edu/bhavnani/wp-content/uploads/2018/04/aa.pdf (accessed
    10 May 2018)
Bouchard, Thomas J. et al. 'Sources of Human Psychological
    Differences: The Minnesota Study of Twins Reared Apart'.
    Science 250, no. 4978 (12 October 1990), 223–8
Bryc, Katarzyna et al. 'The Genetic Ancestry of African
    Americans, Latinos, and European Americans across the
    United States'. *American Journal of Human Genetics* 96, no. 1
    (January 2015), 37–53
Chen, Jieming et al. 'Genetic Structure of the Han Chinese

Population Revealed by Genome-wide SNP Variation.'
American Journal of Human Genetics 85, no. 6 (December
2009), 775–85

Coffey, Diane et al. 'Explicit Prejudice'. *Economic and Political
Weekly* 53, no. 1 (6 January 2018)

Colman, Andrew. 'Race Differences in IQ: Hans Eysenck's
Contribution to the Debate in the Light of Subsequent
Research'. *Personality and Individual Differences* 103,
(September 2016), 182–9

Comfort, Nathaniel. 'Books & Arts: Genetic Determinism
Redux'. *Nature* 561 (27 September 2018), 461–3

Cullinane, Carl and Kirby, Philip. Research Brief: 'Class
Differences: Ethnicity and Disadvantage'. The Sutton Trust,
November 2016, https://www.suttontrust.com/wp-content/
uploads/2016/11/Class-differences-report_References-
available-online.pdf

Dickens, William T. and Flynn, James R. 'Black Americans
Reduce the Racial IQ Gap'. *Psychological Science* 17, no. 10
(1 October 2006), 913–20

'Editorial: Against Bigotry'. *Nature* 548 (17 August 2017), 259

Edson, Lee. 'Jensenism: The Theory that I.Q. is Largely
Determined by the Genes'. *New York Times Magazine* (31
August 1969), 10

Einstein, Albert. *The Travel Diaries of Albert Einstein: The Far
East, Palestine, and Spain, 1922–1923*, ed. Ze'ev Rosenkranz.
Princeton, NJ: Princeton University Press, 2018

'Father Kills Daughter for Marrying Outside their Caste in
Maharashtra', *Hindustan Times*, 6 April 2017, http://www.
hindustantimes.com/india-news/father-allegedly-kills-daughter-
for-marrying-outside-their-caste-in-maharashtra/story-WKLKo
54zk5IXznDJtThxnI.html (accessed 9 January 2018)

Feldman, Marcus W. and Ramachandran, Sohini. 'Missing
Compared to What? Revisiting Heritability, Genes and
Culture'. *Philosophical Transactions of the Royal Society B*
373, no. 1743 (5 April 2018)

Flynn, James R. 'The Mean IQ of Americans: Massive Gains
     1932 to 1978'. *Psychological Bulletin* 95, no. 1 (1984),
     29–51
Flynn, James R. 'Massive IQ Gains in 14 Nations: What IQ Tests
     Really Measure'. *Psychological Bulletin* 101, no. 2 (1987),
     171–91
Gates-Coon, Rebecca. 'The Children of Sally Hemings'. Library
     of Congress, May 2002, https://www.loc.gov/loc/lcib/0205/
     hemings.html (accessed 11 May 2018)
Hart, Betty and Risley, Todd R. *Meaningful Differences in the
     Everyday Experience of Young American Children*. Baltimore:
     Paul H. Brookes, 1995
Hegewisch, Ariane and Williams-Baron, Emma. 'The Gender
     Wage Gap: 2017 Earnings Differences by Race and Ethnicity'.
     Institute for Women's Policy Research, 7 March 2018, https://
     iwpr.org/publications/gender-wage-gap-2017-race-ethnicity/
     (accessed 10 May 2018)
Holden, Constance. 'Identical Twins Reared Apart'. *Science* 207,
     no. 4437 (21 March 1980), 1323–5 and 1327–8
Human Rights Watch, '"They Say We're Dirty" – Denying an
     Education to India's Marginalized', Report, 22 April 2014,
     https://www.hrw.org/report/2014/04/22/they-say-were-dirty/
     denying-education-indias-marginalized (accessed 9 January
     2018)
Ilardo, Melissa A. et al. 'Physiological and Genetic Adaptations to
     Diving in Sea Nomads', *Cell* 173, no. 3 (April 2018), 569–80
Kochhar, Rakesh and Cilluffo, Anthony. 'How Wealth Inequality
     Has Changed in the U.S. Since the Great Recession, by Race,
     Ethnicity and Income'. Pew Research Center, 1 November
     2017, http://www.pewresearch.org/fact-tank/2017/11/01/
     how- wealth-inequality-has-changed-in-the-u-s-since-the-great-
     recession-by-race-ethnicity-and-income/ (accessed 15
     November 2018)
Moore, Elsie G. 'Family Socialization and the IQ Test
     Performance of Traditionally and Transracially Adopted Black

Children'. *Developmental Psychology* 22, no. 3 (May 1986), 317–26

Nisbett, Richard E. et al. 'Intelligence: New Findings and Theoretical Developments'. *American Psychologist* 67, no. 2 (February–March 2012), 130–59

Plomin, Robert, *Blueprint: How DNA Makes Us Who We Are*, London: Allen Lane, 2018

Plomin, Robert and Deary, Ian J. 'Genetics and Intelligence Differences: Five Special Findings'. *Molecular Psychiatry* 20 (2015), 98–108

Plomin, Robert and von Stumm, Sophie. 'The New Genetics of Intelligence'. *Nature Reviews Genetics* 19 (March 2018), 148–59

Ropers, H.-Hilger and Hamel, Ben C. J. 'X-linked Mental Retardation'. *Nature Reviews Genetics* 6 (January 2005), 46–57

Rushton, J. Philippe and Jensen, Arthur R. 'The Totality of Available Evidence Shows the Race IQ Gap Still Remains'. *Psychological Science* 17, no. 10 (October 2006), 921–2

Schwekendiek, Daniel Jong. 'Height and Weight Differences between North and South Korea'. *Journal of Biosocial Science* 41, no. 1 (January 2009), 51–5

Singh, Rama Shankar. 'The Indian Caste System, Human Diversity and Genetic Determinism', in *Thinking about Evolution: Historical, Philosophical, and Political Perspectives, Volume 2*, ed. Rama Shankar Singh et al. Cambridge: Cambridge University Press, 2000

Snickers, Suzanne et al. 'Genome-Wide Association Meta-analysis Of 78,308 Individuals Identifies New Loci and Genes Influencing Human Intelligence'. *Nature Genetics*. 49, no. 7 (July 2017), 1107–12

Stevens, William K. 'Doctor Foresees an I.Q. Caste System'. *New York Times*, 29 August 1971

Witty, P. A. and Jenkins, M. A. 'Intra-Race Testing and Negro Intelligence'. *Journal of Psychology: Interdisciplinary and Applied* 1 (1936), 179–92

10 Black Pills

Adeyemo, Adebowale et al. 'A Genome-Wide Association Study of Hypertension and Blood Pressure in African Americans'. *PLoS Genetics* 5, no. 7, (17 July 2009)

Arias, Elizabeth et al. 'United States Life Tables'. National Vital Statistics Reports of the United States Centers for Disease Control and Prevention 66, no. 4 (August 2017)

'Asthma and African Americans', US Department of Health and Human Services Office of Minority Health website, https://minorityhealth.hhs.gov/omh/browse.aspx?lvl=4&lvlid=15 (accessed 10 July 2018)

Bailey, Zinzi D. et al. 'Structural Racism and Health Inequities in the USA: Evidence and Interventions. *The Lancet* 389, no. 10077 (8 April 2017), 1453–63

'Cardiovascular disease', NHS, https://www.nhs.uk/conditions/cardiovascular-disease/ (accessed 9 July 2018)

Chae, David H. et al. 'Area Racism and Birth Outcomes among Blacks in the United States. *Social Science & Medicine* 199 (February 2018), 49–55

Cooper, Richard S. 'Race in Biological and Biomedical Research'. *Cold Spring Harbor Perspectives in Medicine* 3, no. 11 (November 2013)

Cooper, Richard S. and Rotimi, Charles. 'Hypertension in Blacks'. *American Journal of Hypertension* 10, no. 7 (1 July 1997), 804–12

Cooper, Richard S. et al. 'An International Comparative Study of Blood Pressure in Populations of European vs. African Descent. *BioMed Central Medicine* 3, no. 2 (5 January 2005)

Cooper, Richard S. et al. 'Elevated Hypertension Risk for African-Origin Populations in Biracial Societies: Modeling the Epidemiologic Transition Study'. *Journal of Hypertension* 33, no. 3 (March 2015), 473–81

Curtin, Philip D. 'The Slavery Hypothesis for Hypertension

among African Americans: the Historical Evidence'. *American Journal of Public Health* 82, no. 12 (December 1992), 1681–6

Dubner, Stephen J. 'Toward a Unified Theory of Black America'. *New York Times Magazine* (20 March 2005)

Faturechi, Robert, 'When a Study Cast Doubt on a Heart Pill, the Drug Company Turned to Tom Price', *ProPublica*, 19 January 2017, https://www.propublica.org/article/when-a-study-cast-doubt-on-heart-pill-the-drug-company turned-to-tom-price (accessed 18 July 2018)

Ferdinand, Keith C. et al. 'Disparities in hypertension and Cardiovascular Disease in Blacks: The Critical Role of Medication Adherence'. *Journal of Clinical Hypertension* 19, no. 10 (October 2017), 1015–24

Garcia, Richard S. 'The Misuse of Race in Medical Diagnosis'. *Pediatrics* 111, no. 5 (May 2004), 1394–5

'Global Health Observatory Data: Raised Blood Pressure', World Health Organization website, http://www.who.int/gho/ncd/risk_factors/blood_pressure_prevalence_text/en/ (accessed 10 July 2018)

Grim, Clarence E. and Robinson, Miguel. 'Commentary: Salt, Slavery and Survival: Hypertension in the African Diaspora'. *Epidemiology* 14, no. 1 (January 2003), 120–2

Hammermeister, Karl E. et al. 'Effectiveness of Hydralazine/Isosorbide Dinitrate in Racial/Ethnic Subgroups with Heart Failure'. *Clinical Therapeutics* 31, no. 3 (March 2009), 632–43

He, Feng J. et al. 'WASH – World Action on Salt and Health'. *Kidney International* 78, no. 8 (2 October 2010), 745–53

'Health A-Z: High blood pressure (hypertension)', National Health Service, https://www.nhs.uk/conditions/high-blood-pressure-hypertension/ (accessed 28 June 2018)

'Hypertension in adults: diagnosis and management: Clinical guideline', National Institute for Health and Care Excellence, published August 2011 and last updated November 2016, https://www.nice.org.uk/guidance/cg127/chapter/1-Guidance

#initiating-and-monitoring-antihypertensive-drug-treatment-including-blood-pressure-targets-2 (accessed 10 July 2018)

Kahn, Jonathan D. 'Race in a Bottle'. *Scientific American* 297, no. 2 (August 2007), 40–5

Kahn, Jonathan D. 'Beyond BiDil: the Expanding Embrace of Race in Biomedical Research and Product Development'. *St. Louis University Journal of Health Law & Policy* 3 (2009), 61–92

Kahn, Jonathan D. 'Revisiting Racial Patents in an Era of Precision Medicine'. *Case Western Reserve Law Review* 67, no. 4 (2017), 1153–69

Kalinowski, Leszek et al. 'Race-specific Differences in Endothelial Function: Predisposition of African Americans to Vascular Diseases'. *Circulation* 109, no. 21, 1 June 2004, 2511–17

Kaufman, Jay S. 'Statistics, Adjusted Statistics, and Maladjusted Statistics'. *American Journal of Law & Medicine* 43 (May 2017), 193–208

Kaufman, Jay S. and Cooper, Richard S. 'Use of Racial and Ethnic Identity in Medical Evaluations and Treatments', in *What's the Use of Race? Modern Governance and the Biology of Difference*, ed. Ian Whitmarsh and David S, Jones. Cambridge, MA: MIT Press, 2010

Kaufman, Jay S. and Hall, Susan A. 'The Slavery Hypertension Hypothesis: Dissemination and Appeal of a Modern Race Theory'. *Epidemiology* 14, no. 1 (January 2003), 111–18

Kaufman, Jay S. et al. 'Socioeconomic Status and Health in Blacks and Whites: The Problem of Residual Confounding and the Resiliency of Race'. *Epidemiology* 8, no. 6 (November 1997), 621–8

Khan, Jawad M. and Beevers, Gareth D. 'Management of Hypertension in Ethnic Minorities'. *Heart* 91, no. 8 (2005), 1105–9

Kidambi, Srividya et al. 'Non-replication Study of a Genome-

Wide Association Study for Hypertension and Blood Pressure in African Americans'. *BioMed Central Medical Genetics* 13, no. 27 (April 2012)

Nelson, Alondra. *The Social Life of DNA: Race, Reparations and Reconciliation after the Genome*. Boston, MA: Beacon Press, 2016

'NIH Policy and Guidelines on The Inclusion of Women and Minorities as Subjects in Clinical Research', United States National Institutes of Health, https://grants.nih.gov/grants/funding/women_min/guidelines.htm (accessed 20 August 2018)

Non, Amy L. et al. 'Education, Genetic Ancestry, and Blood Pressure in African Americans and Whites'. *American Journal of Public Health* 102, no. 8 (August 2012), 1559–65

Nyenhuis, Sharmilee M. et al. 'Race is Associated with Differences in Airway Inflammation in Patients with Asthma'. *Journal of Allergy and Clinical Immunology* 140, no. 1 (January 2017), 257–65

Obasogie, Osagie K. 'Oprah's Unhealthy Mistake'. *Los Angeles Times*, 17 May 2007

Ordúñez, Pedro et al. 'Blacks and Whites in Cuba Have Equal Prevalence of Hypertension: Confirmation from a New Population Survey'. *BioMed Central Public Health* 13, no. 169 (2013)

Roberts, Dorothy. *Fatal Invention: How Science, Politics, and Big Business Re-create Race in the Twenty-first Century*. New York: The New Press, 2012

Sankar, Pamela and Kahn, Jonathan. 'BiDil: Race Medicine or Race Marketing?' *Health Affairs* (July–December 2005), W5-455–63

Smedley, Brian D. et al. *Unequal Treatment: Confronting Racial and Ethnic Disparities in Health Care*. Institute of Medicine (US) Committee on Understanding and Eliminating Racial and Ethnic Disparities in Health Care. Washington, DC: National Academies Press, 2003

Taylor, Anne L. et al. 'Combination of Isosorbide Dinitrate and Hydralazine in Blacks with Heart Failure'. *New England Journal of Medicine* 351, no. 20 (11 November 2004), 2049–57

Tomson, Joseph and Lip, Gregory Y. H. 'Blood Pressure Demographics: Nature or Nurture . . . Genes or Environment?' *BioMed Central Medicine* 3, no. 3 (7 January 2005)

Transparency Market Research, 'Global Cardiovascular Drugs Market: Incessantly Rising Cases of Hypertension and Hyperlipidemia to Fuel Market Growth', 7 March 2018, https://www.prnewswire.com/news-releases/global-cardio vascular-drugs-market-incessantly-rising-cases-of-hypertension-and-hyperlipidemia-to-fuel-market-growth-says-tmr-676097783. html (accessed 17 July 2018)

'U.S. Public Health Service Syphilis Study at Tuskegee', Centers for Disease Control and Prevention, https://www.cdc.gov/tuskegee/index.html (accessed 3 July 2018)

Wolf-Maier, Katharina et al. 'Hypertension Prevalence and Blood Pressure Levels in 6 European Countries, Canada, and the United States'. *Journal of the American Medical Association* 289, no. 18 (14 May 2003), 2363–9

11 The Illusionists

Balter, Michael. 'Brain Man Makes Waves with Claims of Recent Human Evolution'. *Science*, New Series 314, no. 5807 (22 December 2006), 1871, 1873

Balter, Michael. 'Links between Brain Genes, Evolution, and Cognition Challenged'. *Science*, New Series 314, no. 5807 (22 December 2006), 1872

Balter, Michael. 'Schizophrenia's Unyielding Mysteries'. *Scientific American* (May 2017), 55–61

British Medical Association, 'Trend of Growing Numbers of BME Doctors in the Profession Continues', 14 May 2018, https://www.bma.org.uk/about-us/equality-diversity-and-inclusion/equality-lens/trend-2 (accessed 7 June 2018)

Byrd, W. Carson and Ray, Victor E. 'Ultimate Attribution in the Genetic Era: White Support for Genetic Explanations of Racial Difference and Policies'. *Annals of the American Academy of Political and Social Sciences* 661, no. 1 (1 September 2015), 212–35

Cochran, Gregory and Harpending, Henry. *The 10,000 Year Explosion: How Civilization Accelerated Human Evolution.* New York: Basic Books, 2009

Derbyshire. John. 'Evolution of the Brain'. *National Review*, 9 September 2005, https://www.nationalreview.com/corner/evolution-brain-john-derbyshire/ (accessed 6 June 2018)

Evans, Patrick D. et al. 'Microcephalin, a Gene Regulating Brain Size, Continues to Evolve Adaptively in Humans'. *Science*, New Series 309, no. 5741 (9 September 2005), 1717–20

Flynn, James R. *Asian Americans: Achievement beyond IQ.* Hillsdale, NJ: Lawrence Erlbaum, 1991

Fullwiley, Duana. 'Race and Genetics: Attempts to Define the Relationship'. *BioSocieties* 2, no. 2 (2007), 221–37

Hamer, Dean H. and Sirota, Lev. 'Beware the Chopsticks Gene'. *Molecular Psychiatry* 5, no. 1 (February 2000), 11–13

Kanazawa, Satoshi. 'Why Are Black Women Less Physically Attractive than Other Women?' *Psychology Today*, 15 May 2011, available online at http://tishushu.tumblr.com/post/ 5548905092/here-is-the-psychology-today-article-by (accessed 6 June 2018)

Kanazawa, Satoshi and Salahodjaev, Raufhon. 'Why Do Societies with Higher Average Cognitive Ability Have Lower Income Inequality? The Role of Redistributive Policies'. *Journal of Biosocial Science* 50, no. 3 (May 2018), 347–64

Lahn, Bruce and Ebenstein, Lanny. 'Let's Celebrate Human Genetic Diversity'. *Nature* 461, no. 461 (8 October 2009), 726–8

Matthews, Dylan. 'Heritage Study Co-author Opposed Letting in Immigrants with Low IQs'. *Washington Post*, 8 May 2013, https://www.washingtonpost.com/news/wonk/wp/2013/05/08/heritage-study-co-author-opposed letting-in-immigrants-with-low-iqs/?utm_term=.cf882f04806f (accessed 6 June 2018)

Mekel-Bobrov, Nitzan et al. 'Ongoing Adaptive Evolution of
    ASPM, a Brain Size Determinant in Homo sapiens'. *Science*,
    New Series 309, no. 5741 (9 September 2005), 1720–2
Palmié, Stephan. 'Genomics, Divination, "Racecraft"'. *American
    Ethnologist* 34, no. 2 (May 2007), 205–22
Pinto, Rebecca et al. 'Schizophrenia in Black Caribbeans Living
    in the UK: an Exploration of Underlying Causes of the High
    Incidence Rate. *British Journal of General Practice* 58, no. 551
    (June 2008), 429–34
Rothman, Barbara Katz. *The Book of Life: A Personal and Ethical
    Guide to Race, Normality and the Human Gene Study: A
    Personal and Ethical Guide to Race, Normality, and the
    Implications of the Human Genome Project*. Boston: Beacon
    Press, 2001
Rushton, J. Philippe et al. 'No evidence that polymorphisms of
    brain regulator genes Microcephalin and ASPM are associated
    with general mental ability, head circumference or altruism'.
    Biology Letters 3, no. 2 (2007), 157–60
'US migrant children cry for separated parents on audio', BBC
    News, 19 June 2018, https://www.bbc.co.uk/news/world-us-
    canada-44531187 (accessed 19 June 2018)
Varese, Filippo et al. 'Childhood Adversities Increase the Risk of
    Psychosis: A Meta-analysis of Patient-Control, Prospective-
    and Cross-sectional Cohort Studies'. *Schizophrenia Bulletin*
    38, no. 4 (18 June 2012), 661–71
von Verschuer, Otmar. *Racial Biology of the Jews*, trans. from
    German by Charles E. Weber, Ph.D. Reedy, WV: Liberty
    Bell, 1983
Watkins, Eli and Phillip, Abby. 'Trump decries immigrants from
    "shithole countries" coming to US'. CNN Politics, 12 January
    2018, https://edition.cnn.com/2018/01/11/politics/immigrants-
    shithole-countries-trump/index.html (accessed 14 June 2018)
Wicherts, Jelte M. et al. 'The Dangers of Unsystematic Selection
    Methods and the Representativeness of 46 Samples of African
    Test-takers. *Intelligence* 38 (2010), 30–7

Afterword

Bremmer, Ian. 'These 5 Countries Show How the European Far-Right Is Growing in Power'. *Time*, 13 September 2018, http://time.com/5395444/europe-far-right-italy-salvini-sweden-france-germany/ (accessed 20 November 2018)

Haas, Ernst B. 'Review: What is Nationalism and Why Should We Study it?' *International Organization* 40, no. 3 (Summer 1986), 707–44

'Poland Independence: Huge Crowds March Amid Far-right Row', BBC News, 11 November 2018, https://www.bbc.co.uk/news/world-europe-46172662 (accessed 20 November 2018)

Porter, Tom. 'Charlottesville's Alt-right Leaders Have a Passion for Vladimir Putin'. *Newsweek*, 16 August 2017, https://www.newsweek.com/leaders-charlottesvilles-alt-right-protest-all-have-ties-russian-fascist-651384 (accessed 21 November 2018)

Sommer, Allison Kaplan. 'The Global Anti-globalist: Steve Bannon Comes Out as Proud "Racist" on His European Comeback Tour'. *Haaretz*, 11 March 2018, https://www.haaretz.com/us-news/.premium-on-european-tour-steve-bannon-comes-out-as-a-proud-racist-1.5890885 (accessed 20 November 2018)

Stille, Alexander. 'How Matteo Salvini Pulled Italy to the Far Right'. *The Guardian*, 9 August 2018, https://www.theguardian.com/news/2018/aug/09/how-matteo-salvini-pulled-italy-to-the-far-right (accessed 20 November 2018)

Index